Babak Akhgar • Petra Saskia Bayerl
George Leventakis
Editors

Social Media Strategy in Policing

From Cultural Intelligence to Community Policing

Springer

Editors
Babak Akhgar
CENTRIC (Centre of Excellence
in Terrorism, Resilience, Intelligence
and Organised Crime Research)
Sheffield Hallam University
Sheffield, UK

Petra Saskia Bayerl
CENTRIC (Centre of Excellence
in Terrorism, Resilience, Intelligence
and Organised Crime Research)
Sheffield Hallam University
Sheffield, UK

George Leventakis
Civil Emergency Planning
and Civil Protection
Hellenic Parliament
Athens, Greece

ISSN 2523-8507 ISSN 2523-8515 (electronic)
Security Informatics and Law Enforcement
ISBN 978-3-030-22001-3 ISBN 978-3-030-22002-0 (eBook)
https://doi.org/10.1007/978-3-030-22002-0

This Springer imprint is published by the registered company Springer Nature Switzerland AG
The registered company address is: Gewerbestrasse 11, 6330 Cham, Switzerland

Security Informatics and Law Enforcement

Series editor

Babak Akhgar, CENTRIC
(Centre of Excellence in Terrorism, Resilience,
Intelligence and Organised Crime Research)
Sheffield Hallam University
Sheffield, UK

Editor's Note:

The primary objective of this book series is to explore contemporary issues related to law enforcement agencies, security services and industries dealing with security related challenges (e.g., government organizations, financial sector insurance companies and internet service providers) from an engineering and computer science perspective. Each book in the series provides a handbook style practical guide to one of the following security challenges:

Cyber Crime - Focuses on new and evolving forms of crimes. Books describe the current status of cybercrime and cyber terrorism developments, security requirements and practices.

Big Data Analytics, Situational Awareness and OSINT- Provides unique insight for computer scientists as well as practitioners in security and policing domains on big data possibilities and challenges for the security domain, current and best practices as well as recommendations.

Serious Games – Provides an introduction into the use of serious games for training in the security domain, including advise for designers/programmers, trainers and strategic decision makers.

Social Media in Crisis Management – explores how social media enables citizens to empower themselves during a crisis, from terrorism, public disorder, and natural disasters

Law enforcement, Counterterrorism, and Anti-Trafficking – Presents tools from those designing the computing and engineering techniques, architecture or policies related to applications confronting radicalisation, terrorism, and trafficking.

The books pertain to engineers working in law enforcement and researchers who are researching on capabilities of LEAs, though the series is truly multidisciplinary – each book will have hard core computer science, application of ICT in security and security / policing domain chapters. The books strike a balance between theory and practice.

More information about this series at
http://www.springer.com/series/15902

Dedicated to Aryan H. Akhgar *You are always in our hearts*

Preface

Context-sensitive and timely intelligence sharing is the cornerstone of modern-day policing. Social media can be a rich source of data that, with relevant and appropriate processing and analysis, can turn into actionable intelligence. The relevance of social media is hard to overstate, now they have become an integral part of our communication. More people today receive their news over social media, for instance, than over traditional news outlets.

In line with their increasing relevance, police forces across the world have established a firm foot in social media—be it in global networks such as Twitter, YouTube or Facebook or other more local variations. These platforms offer efficient access to citizens and have been credited with improving police-citizen relations as well as helping in preventing and solving crimes. At the same time, effects of 'filter bubbles', 'fake news' and polarisation become increasingly apparent.

Whether social media are 'a force for good or bad' is not a question we can and want to answer. In contrast, in which way police forces can and should use social media and how they can support police work is a question that interests us both as professionals and as citizens.

Our motivation for this book is to highlight the role of social media for policing today. Our authors do so—deliberately—from a wide variety of angles: conceptually and strategically as well as in terms of technologies and specific areas of applications.

Social media phenomena have become an integral part of our societies. Police forces thus need to adapt their strategies to embrace and exploit the information on social media platforms for the safety and security of citizens. This book is driven by the belief that police forces' social media strategies need to address the 'softer issues' (such as accounting for a new orientation towards citizens—now also referred to as 'customers' or 'clients', addressing increasingly culturally diverse communities or handling increasing requests for online anonymity) as well as the 'harder issues' of technologies, analytical methods and applications. Social media moreover confront police forces with new challenges such as digital vigilantism. Vigilantism is not a new phenomenon, but social media have certainly provided additional powerful tools for citizens to 'take justice into their own hands'.

Our authors provide hands-on advice and describe concrete applications for a diverse set of law enforcement issues: from crime prevention, identification and reporting to community policing and citizen engagement. By themselves or in combination with other datasets, social media provide law enforcement with rich sources of information, as well as additional channels to engage with citizens. Social media are no longer 'new media', and their strategic as well as practical value for law enforcement is undoubted.

We hope that this book will provide our readers with both the conceptual and practical knowledge and the inspiration to employ social media in new and effective ways.

Sheffield, UK — Babak Akhgar
Sheffield, UK — Petra Saskia Bayerl
Athens, Greece — George Leventakis

Acknowledgements

The editors would like to thank the experts who have contributed to this book, sharing their knowledge, experience and latest research. Our gratitude is also extended to the following organizations and projects:

- CENTRIC (Centre of Excellence in Terrorism, Resilience, Intelligence and Organised Crime Research), UK
- EU-H2020 project INSPEC^{2}T
- EU-H2020 project UNITY
- The European Commission H2020 Research Program
- European Commission DG Home
- Program Committee and Organisers of the 2017 Next Generation in Community Policing (NGCP) Conference

Contents

Contributors

Babak Akhgar CENTRIC, Sheffield Hallam University, Sheffield, UK

Sebastian Allertseder University of Applied Sciences for Public Administration, Fuerstenfeldbruck, Bavaria, Germany

Aitor Álvarez Vicomtech, San Sebastian, Spain

Nerea Aranjuelo Vicomtech, San Sebastian, Spain

Haritz Arzelus Vicomtech, San Sebastian, Spain

Petra Saskia Bayerl CENTRIC, Sheffield Hallam University, Sheffield, UK

Ben Brewster CENTRIC, Sheffield Hallam University, Sheffield, UK

Elisavet Charalambous Advanced Integrated Technology Solutions & Services Ltd, Nicosia, Cyprus

X. Cheng Middlesex University, London, UK

Anastasios Doulamis National Technical University of Athens, Athens, Greece

Nikolaos Doulamis National Technical University of Athens, Athens, Greece

Abdi Fidow Middlesex University, London, UK

Jorge García Vicomtech, San Sebastian, Spain

Helen Grantham CENTRIC, Sheffield Hallam University, Sheffield, UK

Ahmed Hassan Middlesex University, London, UK

Jarmo Houtsonen Police University College, Tampere, Finland

Mahamed Iman Middlesex University, London, UK

Pirjo Jukarainen Police University College, Tampere, Finland

Dimitris Katsaros EXUS Software Ltd., London, UK

Joachim Kersten German Police University, Münster, Germany

George Kokkinis Center for Security Studies (KEMEA), Athens, Greece

Antonis Kostaridis SATWAYS Ltd., Athens, Greece

Nikolaos Koutras Advanced Integrated Technology Solutions & Services Ltd, Nicosia, Cyprus

Fotis Kouretas IMC Information Management & Communications S.A., Athens, Greece

Reinhard Kreissl VICESSE, Vienna, Austria

Olavi Kujanpää Police University College, Tampere, Finland

Norbert Leonardmaier VICESSE, Vienna, Austria

Peter Leškovský Vicomtech, San Sebastian, Spain

George Leventakis Civil Emergency Planning and Civil Protection, Hellenic Parliament, Athens, Greece

Antonios Litke Infili Technologies PC, Zografou, Greece

Anja Louis Languages and Cultures, Sheffield Hallam University, Sheffield, UK

Laurence Marzell Serco Europe, Hampshire, UK

Dimitris Miltiadou INTRASOFT International S.A., Athens, Greece

Holger Nitsch University of Applied Sciences for Public Administration, Fuerstenfeldbruck, Bavaria, Germany

Nikolaos Papadakis National Technical University of Athens, Athens, Greece

Hellenic Army Academy, Vari, Greece

Leonidas Perlepes SATWAYS Ltd., Athens, Greece

M. Petridis Middlesex University, London, UK

Santiago Prieto Vicomtech, San Sebastian, Spain

Eftychios Protopapadakis National Technical University of Athens, Athens, Greece

Aratz Puerto Vicomtech, San Sebastian, Spain

Kari Pylväs Police University College, Tampere, Finland

Nadeem Qazi Department of Mechanical, Aerospace, and Civil Engineering, College of Engineering, Design and Physical Sciences, Brunel University, London, UK

Alice Raven CENTRIC, Sheffield Hallam University, Sheffield, UK

Frauke Reichl Westfälische Wilhelms-Universität Münster, Münster, Germany

Kam Star PlayGen, London, UK

Clifford Sule Middlesex University, London, UK

Jari Taponen Helsinki Police Department, Helsinki, Finland

Sofia Tsekeridou INTRASOFT International S.A., Athens, Greece

Luis Unzueta Vicomtech, San Sebastian, Spain

David Waddington Communication and Computing Research Centre, Sheffield Hallam University, Sheffield, UK

Felix Wentworth PlayGen, London, UK

B. L. William Wong Middlesex University, London, UK

About the Editors

Babak Akhgar is Professor of Informatics and Director of CENTRIC (Centre of Excellence in Terrorism, Resilience, Intelligence and Organized Crime Research) at Sheffield Hallam University, UK, and Fellow of the British Computer Society. He has more than 120 refereed publications in international journals and conferences on strategic information systems with specific focus on knowledge management (KM) and intelligence management. He is member of editorial boards of several international journals and has acted as Chair and Program Committee Member for numerous international conferences. He has extensive and hands-on experience in the development, management and execution of KM projects and large international security initiatives (e.g. the application of social media in crisis management, intelligence-based combating of terrorism and organized crime, gun crime, cyber-crime and cyber terrorism and cross-cultural ideology polarization). In addition to this, he acts as technical lead in EU security projects (e.g. the EU H2020-project TENSOR on dark web). He has co-edited several books on intelligence management. His recent books are titled *Knowledge Driven Frameworks for Combating Terrorism and Organised Crime*, *Emerging Trends in ICT Security*, *Application of Big Data for National Security* and *OSINT for Law Enforcement Agencies.* Professor Akhgar is Board Member of the European Organisation for Security (EOS) and member of the academic advisory board of SAS UK.

Petra Saskia Bayerl is Professor of Digital Communication and Security at CENTRIC (Centre of Excellence in Terrorism, Resilience, Intelligence and Organized Crime Research) at Sheffield Hallam University, UK. She is a regular speaker at police and security conferences and workshops and member of advisory boards of EU projects as well as program Committee Member for international conferences. Her research lays at the intersection of human-computer interaction, organizational communication and organizational change with a special focus on the impact of technological innovations and public safety. Her research has been published in journals such as *MIS Quarterly*, *Communications of the ACM*, *New Media and Society* and *Journal of Organizational Behaviour* as well as international conferences in psychology, management, computational linguistics and computer sciences and books. She has co-edited several books, most recently *Application of Big Data for National Security* (Elsevier), *Open Source Intelligence Investigation: From Strategy to Implementation* (Springer), *Community Policing – A European Perspective* (Springer) and *Digitale Polizeiarbeit* (Springer VS).

George Leventakis, PhD, MBA, MSc is a qualified security expert. He holds a PhD in the area of Risk Assessment Modelling in Critical Infrastructure (CI) Protection, an MBA and an MSc in Risk Management. He has 24 years of professional experience in the public sector, of which 18 years are in security management. He has participated in several national, European and international projects and initiatives, regarding physical security of critical infrastructures, border management (land and sea border surveillance) and civil protection/homeland security technology and operations. He has also participated in tender procedures for complex security systems, including command and control and decision support systems. His research interests include social media platforms in community policing, risk assessment modelling in CI protection, smart borders applications and tools and integrated border management solutions. Since 2006, he was the scientific coordinator of the Center for Security Studies—the Scientific, Advisory and Research Center of the Hellenic Ministry of Interior (KEMEA)—and participated in various European programs funded by the European Commission. He has participated as Senior Researcher in more than 45 EU research projects, has authored several academic papers published in relevant journals and has presented them at academic conferences. Dr. Leventakis has worked and collaborated with many public safety and security agencies in Greece and abroad; from the execution phase of the Risk Management Program for the Sydney 2000 Olympic Games, the Security Program of the Athens 2004 Olympic Games, till more recently on the design and development of National Scale Table Top and Operational Readiness Exercises. He has been involved in the development of threat assessment and vulnerability assessment studies, operational security plans and emergency response plans and procedures for the protection of vital infrastructures and governmental buildings in Greece and EU. Currently he has been assigned as Head of Division at the Civil Emergency Planning and Civil Protection division, within the Hellenic Parliament.

PART I

Social Media and Policing: An Introduction

CHAPTER 1

Introduction: The Police and Social Media

David Waddington

Early Applications

The chapters in this volume reflect a shared interest among the contributing authors in the police use of social media. The nature of this interest relates either to the police prevention or investigation of various forms of crime and disorder, or to the ways in which the police are striving to develop and maintain a more engaging and mutually supportive basis of working in tandem with the public.

Kaplan and Haenlein (2010, p. 61) have usefully referred to social media as "a group of Internet-based applications that build on the ideological and technological foundations of Web 2.0, and that allow the creation and exchange of User Generated Content." One fundamentally key characteristic of such media (e.g. Facebook, Twitter and YouTube) is that they not only enable people to share information with other users, in the form of texts, photographs, videos, etc., but also allow those concerned to engage interactively with one another (The Police Foundation, 2014).

D. Waddington (✉)
Communication and Computing Research Centre,
Sheffield Hallam University, Sheffield, UK
e-mail: d.p.waddington@shu.ac.uk

B. Akhgar et al. (eds.), *Social Media Strategy in Policing*,
Security Informatics and Law Enforcement,
https://doi.org/10.1007/978-3-030-22002-0_1

Common sense dictates that the increasing pervasiveness of social media in all developed societies carries inevitable consequences for day-to-day policing:

> *Not being on social media is not an option for police forces—and this is not only because of the proliferation of new forms of crimes such as financial fraud, identity theft, recruiting for terrorism or the grooming of children by paedophiles in online fora and games. Also, service expectations of citizens are changing. The public increasingly expects faster reactions and the continuous availability of public services independent of physical location.* (Bayerl et al. 2014, p. 1)

Seen from a more positive perspective, social media may also be regarded as a means of promoting greater levels of trust and co-operation between the police and the civilian population, rendering the former far more effective in their work (Bullock 2018; The Police Foundation 2014).

It has been pointed out by Crump (2011) that some UK police forces had already started using social media, albeit tentatively at first, as long ago as 2008. This experimental phase in its gradual uptake was largely instigated by enthusiastic individual officers, working invariably in the absence of any noticeable institutional encouragement and support. Its subsequent rate of growth, both in Europe and the USA, was sluggish to say the least. It was also apt to be employed in a decidedly conservative and unadventurous manner.

This general state of affairs is exemplified by one study of the use of Twitter by American police forces which observed how, with a few notable exceptions, departments were resorting to this medium "primarily to distribute information about crimes and incidents followed by sending information about department happenings, traffic, events, and person identification" and that they "tend to not use the conversational aspects of Twitter including retweeting, replying, or mentioning other Twitter users" (Heverin and Zach 2011, p. 6).

Corresponding research on the adoption of social media by European police forces suggested that, while significant progress was being made in a handful of countries such as the British and Dutch police forces, there was little evidence to suggest that the use of Twitter, Facebook, etc., had become a routine feature of everyday police work across the continent as a whole (Denef et al. 2012). This relatively strong presence of social media within UK police forces was almost certainly a reflection of the massive impetus provided by the publication of the National Police Improvement

Agency's (NPIA's) "*Engage: Digital and Social Media for the Police Service*" report (National Policing Improvement Agency (NPIA) 2010), which advocated the more concentrated and sustained employment of social media as a strategy for keeping citizens much better informed and encouraging greater support for partnership policing initiatives.

A systematic evaluation of the impact of the "Engage" strategy by Crump (2011) revealed that, even by October 2010, some 30 police forces had established a corporate Twitter account. Within the forces concerned, no fewer than 140 neighbourhood and local policing teams had also created separate accounts. The overall take-up pattern was by no means uniform. For example, the Metropolitan Police Service (MPS)—a force consisting of a quarter of all the UK's serving police officers—had failed to exhibit the type of gusto displayed by (say) the Greater Manchester Police in comprehensively establishing Twitter accounts at grass roots level. A generally positive picture had undoubtedly emerged.

Nevertheless, the evidence examined here suggests that, for all its apparent novelty in a hierarchical organisation, Twitter has been most effectively used in support of the traditional roles of the police as a source of appeals for information and a broadcaster of announcements. By the end of 2010, Twitter was already showing its value as a way of getting timely and localised messages out to the public and of adding to the reach of traditional appeals for information. There was less evidence, though, of it forming the basis of the kind of dialogue that could truly be called *engagement* (Crump 2011, p. 24).

A far-reaching briefing report, compiled by the UK Police Foundation, continued to extol the importance of social media beyond the pivotal year of 2008. Its greatest value was seen to reside in its relationship to three crucial, overlapping areas of police activity: (1) providing information to the public in a cheap and timely manner, (2) "providing the police with a way of connecting and building relationships [i.e. engaging with] local communities and members of the public" and (3) improving the gathering and usage of intelligence by "allowing the police to listen to what their communities are saying and to build evidence for investigations by monitoring social media content" (The Police Foundation 2014, p. 3).

The forthcoming chapters of this volume have been written with the intention of highlighting more recent developments in the application of social media across the three areas referred to in the Police Foundation report. In light of the specific topics covered by our contributors, we have chosen to re-categorise the relevant areas of police activity as *investigation and prevention*, and *community inclusion and engagement.*

The particular applications to be featured in this collection are undoubtedly varied and wide-ranging. We therefore aim to establish a preliminary baseline and sense of direction for ensuing discussion by devoting the next two sections of this chapter to illustrative examples of ways that social media has been utilised in the policing of public disorder—an area of particular academic interest to members of the editorial team (e.g. Akhgar et al. 2013; 2017; Denef et al. 2013; Moran and Waddington 2016). In keeping with our chosen structure, the sections will deal firstly with the exemplary police usage of social media to investigate and prevent outbreaks of disorder, followed by attempts by the police to include and involve members of the public in the maintenance or re-establishment of order. A penultimate section will then dwell briefly on a number of practical and ethical considerations arising from these discussions which are likely to be revisited later in the volume. A final section will outline the scope and objectives of the remaining chapters of the book.

Investigation and Prevention

The Police Foundation report rightly maintains that the phenomenal speed with which social media has the capacity to be used in appealing for witnesses or generating other types of relevant information constitutes an extremely "crucial advantage" in terms of intelligence-gathering (The Police Foundation 2014, p. 6). The veracity of this assertion is clearly demonstrated with reference to events surrounding a play-off tie in the 2011 Canadian Stanley Cup hockey tournament which, according to Trottier (2012 p. 412), "was a turning point for social media policing."

The match in question took place in Vancouver on 16th June, where the local team was defeated by Boston. In the wake of this particular contest, some 100,000 people assembled in the centre of the city and proceeded to set fire to cars and vandalise shop windows in the area. Within an hour of this trouble starting, thousands of local citizens logged on to Facebook with the intention of not only registering their outrage but also posting photographic and video evidence (mostly culled from mobile phones) in the hope of bringing those featured to justice. As Trottier (2012, p. 416) explains,

> *These riots are not a novelty in Canada, but they have always had a degree of anonymity; they were never this visible. Yet in 2011, they yielded an unprecedented amount of social media content. Riot-themed groups grew on Facebook,*

and one entitled "Vancouver Riot Pics: Post our Photos" garnered over 100,000 users, over five million views, and countless photographs in under five days. Its members believed that this kind of shaming through visibility is "as strong as a deterrent you will find to prevent this [riot] from happening again" ... While the admissibility of many of these photographs was questionable, the group marked a shift toward greater policing of social life through social media and mobile technologies. Users directly contributed photographs, names, and descriptions of incidents. Police investigations certainly benefitted from the thousands of photographs and hundreds of videos that they received.

A similar approach to this was utilised later that same year by the UK's Metropolitan Police Service (MPS), as part of their quest to bring miscreants involved in the so-called London riots to justice. As part of their initial attempt to identify suspects via crowd sourcing, the MPS posted CCTV and photographic images of the individuals concerned on the Flickr website, and used Twitter to raise public awareness that such a campaign was now in operation. The force backed up this approach by also appealing to members of the public to send in their own images and videos of the event (Crump 2011). Within a day of having been posted, the Flickr image had been retweeted no fewer than 8500 times and viewed on 4.3 million occasions, providing ample proof of the immense potential value of social media for police intelligence gathering (The Police Foundation 2014).

Trottier (2012 p. 416) reveals that although a subsequently commissioned report on the police handling of the Vancouver disorder applauded the way in which they had used social media to harvest the post-riot information referred to above, the authors nevertheless maintained "that Vancouver police did a suboptimal job of *anticipating* rioting based on social media activity, and recommended more aggressive intelligence gathering on social media" (Trottier 2012, p. 416, emphasis added).

In a later article of equal relevance to this chapter, Trottier (2015) outlines the fundamental importance of Open Source Intelligence (OSINT), and the ways in which "investigative agencies" such as the police, are able to gather and interpret data that are technically available to anyone who might choose to access them (Trottier 2015). According to Trottier (2015, p. 533),

OSINT refers to the repurposing of public records for intelligence and investigations, including social media content not protected by privacy settings. While the investigating agent has to manually specify the search criteria and range, tools such as search engines and web crawlers will then automatically retrieve these data and as such are key elements in a process of constructing actionable intelligence from public records. These tools can be used for both retrospective as well as real-time searching. As social media platforms maintain a presence in social life, their users continue to submit information, much of which is publicly accessible by default. Data analytics by law enforcement are located within a context of repurposing social information.

Williams et al. (2013) emphasise the potential value of OSINT in helping the police both to anticipate and, hopefully, to offset the occurrence of disorder. They begin their discussion with reference to an earlier study by King and Waddington (2004) of community tensions arising in the prelude to the 2001 riot in the UK textile-manufacturing town of Burnley in East Lancashire. King and Waddington had shown in their analysis how the pre-riot intelligence gathered in by the Lancashire police (on the basis of anecdotal information from local officers and consultations with neighbourhood forums) had left them with the misguided impression that growing signs of tension and disorder occurring in predominantly Asian areas were symptomatic of an ongoing "turf war" involving rival groups of drugs traders.

It is probable that had their intelligence been more sophisticated and reliable, the police may well have realised that the attitudes and behaviour of the Asian community as a whole reflected fear and resentment caused by the increasing presence and activities of the far right. Williams et al. (2013, pp. 462–463) are justified in speculating that

Had the welter of alternative accounts that are facilitated by social media been available in 2001, it is at least plausible to argue that tension arising out of extreme right-wing political agitation would have been reflected in social media communications thereby indicating other dynamics driving, indeed anticipating, the escalation of this civil unrest.

Acting in the wake not only of the 2001 riots but also of the subsequent English disorders of 2011, Williams et al. set out to show the feasibility of building a "tension engine", capable of predicting imminent communal violence on the basis of ongoing social media activity. In order to accomplish this task, an expert team of researchers was duly assembled, comprising experts in criminology, the sociology of language and computer science, who were given the remit of designing "a social media tension-monitoring system that drew on sociological concepts that could be digitally codified to operate in an online environment" (Williams et al. 2013, p. 470).

This process of "Collaborative Algorithm Design" involved adapting the coding methods used by Harvey Sacks (e.g. Sacks 1972) in his research on *Conversation Analysis and Membership Categorisation Analysis* in the automated collection and analysis of social media discussions appearing on over 100 local community forums in Cardiff and Manchester. These discussions tended to focus on such issues as local parking and burglaries. Nevertheless, by extrapolating from the findings of this research, the team felt confident enough to maintain that "A 'social-listening' of these and individual channels of communication can provide a rich source of data on rising and lowering community tensions adding an additional digital layer onto more conventional terrestrial intelligence creating a form of 'neighbourhood informatics'" (Williams et al. 2013, pp. 476–477).

A more concrete insight—this time into the ways that recent advances in OSINT which have been used in the prediction and deterrence of disorder during protests and demonstrations—is provided by (Dencik et al. 2018). These authors explain how, in recent years, the police remit for handling events of this nature has been handed over to the National Domestic Extremism and Disorder Intelligence Unit (NDEDIU). Starting in 2012, the NDEDIU has included a team of some 17 staff assigned to Social Media intelligence (SOCMINT). Dencik et al. emphasise that the NDEDIU tends to use a variety of "off-the-shelf" OSINT tools and programmes, rather than depending on in-house software developers and engineers. Moreover, those police officers involved tend not to have any knowledge of, or direct input into, the algorithms underlying whatever software they may be applying.

Specialist officers interviewed as part of the study described the main purpose of monitoring social behaviour as improving "situation awareness, in such a way as to identify the potential for disruption and violence at forthcoming protests and demonstrations, and ideally pinpoint any individuals or groups who might be intent on creating trouble of this nature." In practice,

> *Keyword searching is the most dominant practice. Large data sets relating to a particular event are filtered by a list of keywords in order to search for potential threats. "Threats" in this context would be for example particular words associated with violence or disruption ("threat words") and would be followed by an assessment as to whether further action is needed to identify individuals. Lists of keywords and threat words are context specific and dif-*

> *ferent lists are developed depending on the nature of the event the location and the people it is likely to attract (particularly to include sensibilities of language and dual meaning words for example "flared trousers as opposed to a flare being set off"* … *As such algorithms are used to 'filter the noise' in terms of particular words that allow police to assess only highlighted data* (Dencik et al. 2018, p. 1441)

Social media monitoring may also be used to enable police to roughly ascertain what type of people (and how many of them) are likely to be attending. Interviewees also made special reference to the statements made by "influencers"—that is, individuals who might be notable for their "online reach and following", and not necessarily their direct involvement in the event (Dencik et al. 2018, p. 1441). Information of this nature was deemed particularly useful to the police, "in terms of 'engaging' with such individuals and groups before an event or for identifying potential criminal activity resulting from the nature of influencer communication" (Dencik et al. 2018, p. 1442).

Community Inclusion and Engagement

Police attempts to encourage civilian involvement and co-operation in the management of public order have been predicated on the principles of community policing (Waddington 2017). According to Terpstra (2009, pp. 65–66) there is widespread academic agreement that community policing may be characterised by five distinguishing elements:

1. A commitment by the police to promoting "their proximity, visibility and approachability to citizens" in order to generate trust in the police and confidence in the integrity whatever information they might impart.
2. A police focus, not merely on crime but also on a wide range of other neighbourhood issues such as nuisance, disorder and feelings of insecurity.
3. The employment of not only reactive but also preventative and proactive measures by police.
4. A police commitment to co-operating with other agencies (both public and private) in pursuit of community policing objectives.

5. A similar undertaking by the police to co-operate with community members in such a way as to promote and support civilian involvement in the management of crime and disorder.

Terpstra subscribes firmly to the view that police attempts to put these ideals into practice have not been particularly wholehearted. He points to evidence which suggests that police enthusiasm only extends to those initiatives likely to provide a source of useful information and/or intelligence. This is simply because, “for the police, this form of citizen participation has the most concrete benefits and it fits in with their routine work” (Terpstra 2009, p. 69). Bullock (2018) reckons that this problem has been compounded by low public participation rates in community-policing initiatives, by the fact that police–community meetings tend to be dominated numerically by the white middle classes, and that such meetings have, in any case, been looked upon principally as a convenient forum for broadcasting information and rubber-stamping predetermined police strategy.

The advent of social media heralded a more optimistic climate in which the diffusion of Web 2.0 platforms and user-generated content was seen as giving rise to greater openness and transparency by the police, and stimulating more widespread and mutually beneficial dialogue and debate between police officers and their civilian constituencies:

> *Moving police communication from the physical to the virtual, social media have been presented as a way of circumnavigating the aforementioned problems and affording a mechanism through which constabularies might engage with citizens and communities and, in so doing, present themselves as open, participative and democratic … Thus, community policing teams are supposed to use social media to reach out to citizens and communities, to engage them in debate and to listen to and act upon their concerns. It has been hoped that social media will provide an efficient technological response to an enduring conundrum in public policing—that of how to stimulate communication between officers and citizens* (Bullock 2018, p. 247).

Our own research and that of other academics has shown how social media platforms have been used by police forces, with growing levels of confidence and assuredness, to engage local communities in such a way as to offset potentially large-scale occurrences of public disorder (cp. Chap. 6).

An early example of this nature took place in January 2010 when Staffordshire Police employed social media (primarily Twitter and Facebook) as part of *Operation Chime*, which they set up in response to a protest by the Far Right *English Defence League* and a counter-demonstration by opposing political groupings. In the run-up to the event, police increased their visibility on relevant sites, using *#policingstoke* as their hashtag. Then, on the actual day of the protest, the police encouraged the setting up of a "live" Twitter feed, hosted by volunteers from "Pits N Pots", an independent news site-cum-blog, featuring political discussions led by local journalists, politicians, community spokespeople and representatives of the police. In the meantime, two police media personnel monitored relevant social media platforms—to seemingly telling effect:

> *Key messages were published to Twitter and Facebook about the policing of the protest throughout the operation and due to the detailed monitoring police responded to questions from members of the public and dispelled rumours about disorder occurring in residential communities … At one stage a user asked if it was ok to use a local railway station. Within minutes Staffordshire Police responded with information and reassurance. This was re-tweeted numerous times. Dozens of comments about the policing operation were also made using Twitter and the media unit made sure they were quickly responded to. Towards the end of the operation YouTube was used to publish a video showing police action and information about the policing operation. This was promoted to those interested in the event using Twitter, Facebook and PitsnPots sites* (National Policing Improvement Agency (NPIA) 2010, p. 14).

We have previously explained in some detail how it was thanks to a similar, purpose-built social media strategy, carefully constructed and carried out by a four-person Social Media Cell in South Yorkshire Police, that there was a virtually trouble-free outcome to the 5000-strong protest occurring outside the Liberal Democrats' Spring Conference in Sheffield in March 2011 (McSeveny and Waddington 2011). Force strategy in this particular instance had focused on using Twitter, Facebook and other local and national discussion forums, both in the lead-up to and during the demonstration, as ways of establishing and maintaining public confidence in the police, engaging with social media communities and potential protesters, and providing "consistent and informed messages" to all interested parties.

It was with these ends in mind that the Senior Media Officer (SMO) in charge of the Social Media Cell endeavoured for several months prior to the demonstration to build up as sizeable a personal Twitter following as possible. Then, with the demonstration a matter of only 4 weeks away, the SMO began tweeting a much greater quantity of messages of direct relevance to the forthcoming protest. Though often still informative, such messages were primarily designed to establish an upbeat and reassuring "tone" for the eventual policing of the protest—emphasising, for example, that SYP was totally committed to facilitating a safely conducted and meaningful protest.

On the day of the conference itself, the SMO and her colleagues continued to pump out Twitter feeds, with the aim not only of propagating a positive ethos but also to allay insecurities and dispel any rumours that came to light in the course of their ongoing dialogue with the public. Thus, to give one such example, the SMO acted in all due haste to counteract one increasingly popular allegation that SYP were intending to use a gated barrier at the end of the protest site in order to "kettle" (i.e. box in and contain) those gathered in the area. It soon became evident from the content of resulting dialogue that the SMO had succeeded both in persuading her followers that "South Yorkshire Police does not acknowledge kettling as an approved Home Office method of crowd control", and in reassuring them that the gate was being used with the exclusive intention of avoiding potentially dangerous crushing or overcrowding. A similar social media intervention by the SMO also served to scotch an increasingly pervasive and pernicious rumour that police personnel on top of an overlooking city-centre department store were actually snipers with rifles at the ready. It is a measure of the success of the social media strategy that only one arrest occurred in the course of the entire proceedings.

Our subsequent research and publications have focused on the extent to which social media has already been (or might feasibly be) employed with the aims of preventing or curtailing the occurrence of urban riots. One significant conclusion emerging from the reviews and analyses by Moran and Waddington (2016) of major disorders occurring in such places as Paris, Sydney, Athens, London and Ferguson (Missouri, the USA) is that such disturbances were invariably connected to police failures to use social media to do the following:

1. Dispel rumours and correct misinformation.
2. Reassure and placate the public by listening to and responding to their concerns and grievances and.
3. Wherever necessary, apologise (or at least offer up a reasonable justification) for any contentious actions on their part.

In those cases where social media was used as part of police strategy (most notably in the English riots of 2011) it was applied to contrasting effect. By and large, in the summer of 2011, "British police forces not only saw a tremendous growth in the number of Twitter followers. They also, for the first time, engaged with the public on such a large scale via social media, using Twitter as the main platform" (Denef et al. 2013, p. 3471). Aside from the rioting that occurred across the various districts of the capital, other disorders broke out in places as far afield as Birmingham and Manchester. London's Metropolitan Police Service (MPS) and Greater Manchester Police (GMP) were the two constabularies focused on by Denef et al. with the intention of exploring the nature and effectiveness of their use of Twitter during the riots. As they were careful to explain, the choice of these two forces was governed by the fact that, while the MPS played an extremely central role in the riots, "GMP was less effected by the riots, yet, is known among UK police forces for embracing Twitter and has experimented with its use in campaigns before" (Denef et al. 2013, p. 3472).

There were notable discrepancies in the ways in which the two forces made use of social media. Denef et al. observe how, during the peak period of the rioting (encompassing 4–13 August), MPS posted 132 tweets, compared with the corresponding total of 371 posted by GMP. There was a further contrast in the characteristic styles and content of the messages disseminated by the respective forces: while both undoubtedly employed Twitter with the primary the clear intention of gathering incriminating information (e.g. by posting CCTV images of "offenders" on Flickr and leaving phone numbers and website addresses by which to contact the police with relevant details) the GMP placed a much greater emphasis on reassuring and engaging with the public (Denef et al. 2013).

As in the SYP example referred to above, the GMP adopted an avowedly *personalised* approach to dispelling potentially pernicious rumours and dealing with specific queries. For example, in reacting to an online suggestion that the nationwide riots were spreading into Manchester, the GMP instantly reassured its public in the following manner (Denef et al. 2013, p. 3475):

> *GMP: No disorder or riots in Manchester. Speculation about ongoing riots totally inaccurate. GMP monitoring the situation.*

More personal queries submitted by followers were responded to in similar vein (Denef et al. 2013, p. 3475):

> *USR1: @gmpolice is it true that chaos has started in town, carphone warehouse has been done over already??*
>
> *GMP: @USR1 nothing at the moment follow us and we will let you know if there is anything to report.*

The GMP's obvious commitment to displaying due sensitivity towards the feelings and opinions of its followers extended to a willingness to apologise for and/or retract comments deemed offensive by members of the public. On one occasion, for example, a GMP tweet ("*GMP: Mum-of-two, not involved in disorder, jailed for FIVE months for accepting shorts looted from shop. There are no excuses*") provoked a very indignant backlash. In reacting to this message, one particular civilian raised the following objection:

> *The tweet shows enthusiasm, maybe even glee, over the length of the sentence. Particularly with the emphasis of "FIVE months" and "There are no excuses!" It is not the place of the police to comment on, recommend or celebrate the length of a sentence or the defence used in court … [the police] should remain detached and professional when it comes to presenting information to the public* (cited in Denef et al. 2013, p. 3477).

The GMP's response exuded due contrition (Denef et al. 2013, p. 7):

> *GMP: Apologies for any offence caused from last tweet. Comment was not directed at individual person.*
>
> *GMP: Thanks to all for feedback messages—all your comments have been noted. You are right, it is not our place to comment on sentences.*
>
> *GMP: appreciate all feedback. Changing tack slightly—we really want to know what you think we've got right or wrong this week on Twitter*

Within an hour of posting this message, they had deleted the offensive tweet.

It may be taken as indicative of the relative effectiveness and success of the contrasting social media policies adopted by the MPS and GMP that, while the former increased their number of Twitter followers from 4000 to 42,000 with the onset of the riots, the GMP more than quadrupled their following by going from 23,000 to 100,000 (Denef et al. 2013). Denef et al. observe how the GMP enjoyed the crucial advantage of having already established a tried-and-tested social media communication strategy well before the rioting broke out. In addition to the main Twitter account referred to by Denef and his colleagues, the force operated a further 60 localised accounts which were routinely employed to engage in two-way communication with targeted local communities. Procter et al. (2013, p. 435) point out that, since the riots, the MPS appear to have taken a leaf out of the GMP's book and "begun rolling out Twitter to its neighbourhood policing teams, but there is clearly much work to be done."

Some General Issues and Considerations

It is hoped that our brief review, with its deliberately narrow focus on the application of social media in the policing of public disorder, has been sufficient to emphasise the actual and potential benefits to be gained from employing these new technologies in everyday investigative, preventative, and community-oriented (or "partnership") aspects of police work. In highlighting the possible advantages to be derived from embracing these contemporary technological innovations, we have no intention of disregarding the possible obstacles and potential drawbacks associated with their use. Indeed, we consider it important to mention some of the more salient practical and ethical problems alluded to in the literature.

Bullock (2018, p. 248) has recently gone so far as to maintain "that social media are not transforming the nature of routine police–citizen communications", insofar as the relevant new technologies have not been introduced in the ways and extent to which they have been advocated. Basing her assertion on interviews with police social media practitioners, Bullock mainly attributes this situation to the ignorance and disdain of top-ranking police officers who are not only unfamiliar with the relative technology but may also regard it as a time-consuming distraction from what they regard as "real" police work. The fact that there are inadequate numbers of trained officers being specifically assigned to this area is undoubtedly linked to Bullock's observation that "the introduction of social media platforms, which are generally free, open and require light programming, is at odds with the established culture of ICT development within constabularies which tends to be cautious, security conscious and highly risk adverse" (Bullock 2018, p. 251).

Bullock is equally convinced that what she considers to be an overriding police reliance on Twitter (and corresponding reluctance to employ alternative media platforms) is primarily due to the fact that it is looked upon by senior officers as relatively straightforward and not especially time-consuming (Bullock 2018). She cautions that, by acting in this way, "community policing teams may inadvertently be narrowing their reach towards an audience which is dominated by the better educated, better off, younger adult males" (Bullock 2018, p. 253).

If having the requisite numbers of trained staff is crucial to ensuring the effectiveness of police social media systems, then having the corresponding technological capacity may be considered equally important. It became evident during the 2011 riots, for example, that "the systems for developing intelligence from large flows of social media data were not always sufficient to cope with the demand to process it rapidly enough to usefully inform strategies and tactics on the ground" (The Police Foundation 2014, p. 8). It was particularly evident how police personnel experienced great difficulty in being able to ascertain the true accuracy and credibility of the information that was flooding in at great speed and in overwhelming volumes (The Police Foundation 2014). It was little wonder that, in the wake of the riots, a report by Her Majesty's Inspectorate of Constabulary advocated the setting up of a national public order intelligence hub, capable of identifying and analysing changes in community tension levels on the basis of social media (The Police Foundation 2014).

Dencik et al. have also written extensively about some of the possible practical issues involved in the police use of OSINT. One obvious potential drawback in their estimation concerns the fact that "big data" analysis is far from being the "strictly impartial", automated process it is invariably cracked up to be. Indeed, it is important to remember at all times how,

> *The role of human input, both in terms of designing the algorithms as well as any analysis and interpretation of such data, remains central in data-driven governance. The notion that big data may absolve human errors and allow for 'objective' or 'efficient' forms of governance, therefore, is largely mythical in the context of this study at least. Rather, big data is predominantly used to identify patterns that are subjectively (humanly) interpreted and assessed, not least in the identification of any anomalies within these patterns. Thus, discretion (and assumptions and ideology) is a key feature in data-driven policing. In particular, pre-existing knowledge, intelligence and broader societal understandings of events continue to shape and determine big data analyses.* (Dencik et al. 2018, p. 1445)

A closely related factor concerns the undue implicit influence invariably being exerted by a "dominant police procurement model" which dictates that most of the software packages used in OSINT analysis are obtained from commercial organisations with a particular specialism in marketing. It is imperative to bear in mind that such ubiquitous terms as "sentiment" and "influencers" (and the meanings and implications attached to them) were initially coined and applied in marketing analysis, rather than being customised to issues of law enforcement (Dencik et al. 2018, p. 1445). It also requires emphasising that social media users often develop distinctive cultures of expression and interaction which are unique to particular platforms, and which may therefore be open to misinterpretation. Finally, it is just possible that the messages being conveyed are inaccurate, bogus and prone to bravado and exaggeration (Dencik et al. 2018).

Practical problems are one thing. Of "particular concern" to Trottier is the corresponding *ethical* consideration that, whilst it is apparent that sites such as Facebook have become a vital component of police investigation, individual users may be totally unaware of the related processes of "exposure and exploitation" they are currently being subjected to (Trottier 2012, p. 421). It cannot be denied that any online police exercise of this nature will automatically suck in even the most casual, unwitting or "innocent" user of the social media concerned, thus recasting them as a potential "suspect" (Trottier 2015, p. 541). Eventually, the mushrooming possibility will arise that, as users become more and more aware of the potentially "punitive consequences" of saying "incriminating" things online, they will become more circumspect about entering spaces which they know to be "policed through OSINT analysis" (Trottier 2015, p. 542).

Contemplating Ways Forward

The somewhat pessimistic tone accompanying many academic analyses of the police use of social media is not universal. A more optimistic outlook is certainly projected in a recent study by Williams et al. (2018) which focuses on the use of Twitter and Facebook by five major US police departments. These authors detected a notable shift away from the more conventional police social media emphasis on imparting information to the public to one of attempting to engage the relevant civilian populations in two-way communication and, hopefully, establish an increasingly strong rapport:

> *Unlike some prior studies … we observed that social media did facilitate some interactivity, i.e., dialogic communication, between police and citizens, an encouraging sign for those who would employ it for advancing community*

> *policing goals. This finding together with some of our interview data suggest that social media communications can support community policing efforts in ways that cannot be replicated by individual officers in the field, to reach a targeted set of audiences. It requires a relatively low investment of time, money and labor resources … and can reach more people more quickly than face-to-face and many other communications media. Thus, police agenda setting through social media affords opportunities for departments wishing to expand their reach and interaction with their community networks* (Williams et al. 2018, no pages).

It is evident from the preceding section that the potential benefits of this nature will not be realised in the absence of enabling investment. According to the Police Foundation report already referred to above,

> *Deploying sufficient personnel is one such investment. The police need to be able to analyse and manage the information they receive quickly and effectively. This is of particular importance in relation to critical incidents, when there is a need to analyse information at speed. But it takes time to manage information effectively. Setting up systems to mine and scrape social media for intelligence purposes might assist the police in monitoring community tension levels and anticipating behaviour, but it only comes at considerable additional expense* (The Police Foundation 2014, p. 10).

The report acknowledges that further research and practical experimentation is undoubtedly necessary in order to address, and hopefully resolve, the type of practical and ethical issues alluded to above:

> *Work remains to be done on developing systems and strategies that help embed the use of social media into everyday working practices and ensure that lessons are learnt in terms of what constitutes best practice. Research into the public perception of the police's use of social media may be of benefit, particularly if it helped to pin down what the public wants from a police social media account, how they view different types of tweets or posts and what impact social media has on public confidence in the longer-term. With 87 per cent of young people using social media, it represents a crucial form of communication now and in the future and could become a vital element in helping to build trust and confidence and improve policing in years to come.* (The Police Foundation 2014, p. 10)

Comments of this nature are a testament to the importance of the chapters contributed by academics and practitioners to this volume.

References

Akhgar, B., Fortune, D., Hayes, R., Guerra, B. and Manso, M. (2013). Social media in crisis events: Open networks and collaboration supporting disaster response and recovery, *Technologies for Homeland Security (HST), 2013 IEEE International Conference on.* IEEE, 12–14 November 2013, Waltham MA, USA: 760–765.

Akhgar, B., Staniforth, A., & Waddington, D. (Eds.). (2017). *Application of social media in crisis management.* London: Springer.

Bayerl, P.S., Horton, K., Jacobs, G. and Akhgar, B. (2014). Who wants police on social media? In A. Rospigliosi & S. Greener (Eds.), *Proceedings of the European Conference on Social Media (ECSM)*, University of Brighton, Brighton UK, 10–11 July 2014.

Bullock, K. (2018). The police use of social media: Transformation or normalisation? *Social Policy and Society, 17*(2), 245–258.

Crump, J. (2011). What are the police doing on twitter? Social media, the police and the public. *Policy and Internet, 3*(4). article 7.), 1.

Dencik, L., Hintz, A., & Carey, Z. (2018). Prediction, pre-emption and limits to dissent: Social media and big data uses for policing protests in the United Kingdom. *New Media and Society, 20*(4), 1433–1450.

Denef, S., Bayerl, P. S., & Kaptein, N. (2012). Cross-European approaches to social media as a tool for police communication. *European Police Science and Research Bulletin, 6*, 11–14.

Denef, S., Bayerl, P.S. and Kaptein, N. (2013) Social media and the police: Tweeting practices of British police forces during the august 2011 riots. *Paper presented at CHI 2013: Changing Perspectives*, April 27–May 2, Paris, France.

Heverin, Th., & Zach, L. (2011). Twitter for city police department information sharing. *Proceedings of the Association for Information Science and Technology, 47*(1), 1–7.

Kaplan, A. S., & Haenlein, M. (2010). Users of the world unite! The challenges and opportunities of social media. *Business Horizons, 53*, 59–68.

King, M., & Waddington, D. (2004). Coping with disorder? The changing relationship between police public order strategy and practice - a critical analysis of the Burnley riot. *Policing and Society, 14*(2), 118–137.

McSeveny, K., & Waddington, D. (2011). Up close and personal: The interplay between information technology and human agency in the policing of the 2011 Sheffield Anti-Lib Dem protest. In B. Akghar & S. Yates (Eds.), *Intelligence management (knowledge driven frameworks for combating terrorism and organised crime)*. New York: Springer.

Moran, M., & Waddington, D. P. (2016). *Riots: An international comparison.* London: Palgrave Macmillan.

National Policing Improvement Agency (NPIA). (2010). *Engage: Digital and social media engagement for the police service*. London: Author.
Procter, R., Crump, J., Karstedt, S., Voss, A., & Cantijoch, M. (2013). Reading the riots: What were the police doing on twitter? *Policing and Society, 23*(4), 413–436.
Sacks, H. (1972). An initial investigation of the usability of conversational data for doing sociology. In D. N. Sudnow (Ed.), *Studies in social interaction*. New York: Free Press.
Terpstra, I. (2009). Community policing in practice: Ambitions and realization. *Policing, 4*(1), 64–72.
The Police Foundation. (2014). *The briefing: Police use of social media*. London: The Police Foundation.
Trottier, D. (2012). Policing social media. *Canadian Review of Sociology, 49*(4), 411–425.
Trottier, D. (2015). Open source intelligence, social media and law enforcement: Visions, constraints and critiques. *European Journal of Cultural Studies, 18*(4–5), 283–304.
Waddington, D. (2017). Police liaison approaches to managing political protest: A critical analysis of a prominent U.K. example. In P. S. Bayerl, R. Karlovic, B. Akhgar, & G. Markarian (Eds.), *Community policing—A European perspective: Strategies, best practices and guidelines*. London: Springer.
Williams, C. B., Fedorowicz, J., Kavanaugh, A., Mentzer, K., Bennett Thatcher, J., & Xu, J. (2018). Leveraging social media to achieve a community policing agenda. *Government Information Quarterly, 35*, 210. https://doi.org/10.1016/j.giq.2018.03.001
Williams, M. L., Edwards, A., Housley, W., Burnap, P., Rana, O., Morgan, J., et al. (2013). Policing cyber-neighbourhoods: Tension monitoring and social media networks. *Policing and Society, 23*(4), 461–481.

CHAPTER 2

Brave New Apps: An Essay on Community Policing and Social Media

Joachim Kersten, Norbert Leonardmaier, and Reinhard Kreissl

Introduction

Over the last decades, the concept of community policing (CP) has consistently been lacking precise definitions, for example, with respect to what is exactly meant by "community". As Peter Manning (2010, p. 149) has pointed out, "*The matter of undefined 'community' remains a problem for the police, for researchers and for community members.*"

"Community" is loaded with a connotation of pastoral life-worlds, a rural village where families and their offspring act within rather strict sets of rules, norms and values. This depicts the ideal type of *Gemeinschaft* (Tönnies 1887). In contrast to this, the conceptual and practical development of CP in Western societies has originated from the urban territory of

J. Kersten (✉)
German Police University, Münster, Germany

N. Leonardmaier · R. Kreissl
VICESSE, Vienna, Austria

B. Akhgar et al. (eds.), *Social Media Strategy in Policing*,
Security Informatics and Law Enforcement,
https://doi.org/10.1007/978-3-030-22002-0_2

policing, that is, law enforcement in modern cities. Urban lifestyles and modernization are at the basis of *Gesellschaft* (ibid.). More often than not, CP has at its core the idea of a restoration of norms of social control of the *Gemeinschaft* tradition. These were seen to have stabilized human life-worlds that have deteriorated due to modernization, urbanization and migration. CP carries with it the promise that this (often fictitious) model of rules and order will be reconstructed or revitalized through the involvement of police in neighbourhood affairs against disorder, incivilities, conflict and crime.

Common law traditions of countries from where CP concepts have originated, have structures of criminal law and criminal justice very different from civil (Roman) law continental European legal traditions and policing structures. Regularly, EU CP research does not bother much to take this into account.

Real and Virtual Communities

In a similar vein, research has not mustered due diligence concerning differences between "real" and "virtual" communities (Keane 2016, p. 108). Social media exchange as virtual interaction does reveal features of "village talk" (e.g. praise and blame gossip that reinforce stereotypes about the "established" and the "outsiders") (Elias and Scotson 1994). But at its basis we find arbitrariness and non-committal, its "casual nature" (DPZ 2017). Also, much more than is the case with face-to-face or even digital email communication, there is often a pathological tendency to excitation and agitation. Whether YouTube, newspaper user comments or presidential Twitter messages, reason, good measure, objectivity, respect for other opinions and decency fly out of the chimney when excited, angry, upset or frightened individuals type messages and make them public. More than any other interaction between persons, social media talk uncovers pathological structures of digital communication. Messages are gossip and may be embarrassing, and often they are meant to hurt and express hate. Often, they are simply stupid. In a word, we might look at a digital heap of crap, unfortunately, with very real consequences of exclusion and stigmatization.

In the light of this, it does not come as a surprise that particularly younger people have started to have second thoughts about their online activities. A vast majority of young users "*would not mind if social media had never been invented*" and "*young people are disillusioned with the negative aspects ..., such as online abuse and fake news*"; schoolchildren say that they have "*taken temporary digital detoxes to escape social media*" (The Guardian 2017).

CP is defined by its local nature. By definition, real communities are bound to locations, and so are related policing activities. Accordingly, definitions vary a lot. The most decisive indicator of implementation of CP is defined as the actual "*collaboration between police officers, residents and organized stakeholders*" (Chappell 2009, p. 11). Obviously, the central focus of most empirical research on CP practices is on big-city police departments, and so there goes the neighbourhood of the *Gemeinschaft* type.

Patrol Officer Resistance and the Release of the "Inner Crime Fighter"

Methodological approaches to CP research mostly rely on quantitative and survey data. This poses a problem, since survey data gauge what police managers/administrators *think* or *want to believe* about CP implementation and effects. Thus, findings contain an inbuilt bias. They tend to systematically conceal the gap between *philosophy*, expressed by police management/administrator attitudes, and patrol officer/street line actual implementation of CP as an *activity* (Brainard and Edlins 2014). Patrol officers, even middle management supervisors, may be much less enthusiastic about CP implementation than top management (Chappell 2014).

From direct observational studies it appears that some officers do not see CP activities as a meaningful part of their activities. If such "cultural resistance" against CP extends to the middle management level, a marginal status of CP is likely; "*Even in departments that ostensibly adopt the philosophy department-wide, the extent to which it is part of everyday operations is still questionable*" (Chappell 2014, p. 10).

From such qualitative research the support role of supervisors and the evaluation of officers on CP performance and problem solving appear to be of paramount impact. There ought to be positive rewards for CP activities related to career chances and promotion. Another observation from field research: observers noticed frequent "down time" periods during patrol rides. However, this time was not used for CP activities. Therefore, the "resource problem" cited by so many patrol officers may very well be a perception and not so much a reality.

In US police departments, CP was introduced mainly because of grave police organizational problems, in particular their "inability, to reduce crime rates and improve safety" (Manning 2010). As such the conceptualization of CP can be seen as the counterpart of reactive policing (Corsianos 2012, p. 153). Accordingly, several empirical studies dealing with CP (e.g. Skogan 2006) have demonstrated that officer motivation, or lack thereof ("cultural resistance"; cf. Chappell 2014, p. 18), constitutes one of the key variables for the practical implementation of CP.

Street (patrol) officers' self-image in the USA is often linked to ideologies of hyper-masculinity, namely, that of the "crime-fighter". This appears to be one of the enduring norms of what is generally referred to as "cop-culture" (Corsianos 2012, p. 160). This normative (mythological) imagery is the anathema of the envisaged practice of CP: talking to citizens, being accessible to their worries and concerns, helping them in networks with other agents and neighbourhood activist networks in tidying up (mostly) low-economic status neighbourhoods. These two policing concepts, crime-fighting and community policing, are not compatible. This, however, seems not to be sufficiently clear in the minds of CP software and app engineers. Their models and applications tend to overemphasize the crime-fighting component of police work, not crime reduction or prevention. A particularly questionable case was the "practical demonstration" of app use during the Iraklion NGCP Conference 2017.[1]

For a growing number of social scientists co-operating with software partners in projects as INSPEC^{2}T (see Chapters 12 and 13) or similar ventures presented at the NGCP October 2017 convention, it transpires that the binary logics of software app engineering and the "fuzzy" logics of real police work may not go together so well. Also, the ideology of "real" police work as fighting crime and criminals as frequently envisaged by the software component of technology centred security research has at its basis norms of hyper-masculinity. By definition, for crime fighters community policing represents a not much appreciated world of taking care, tidying up, cleaning and solving conflicts through communication—in a word, an ensemble of social skills that are despised as feminine and soft, and as such seen as non-masculine and not as "real cop's work" (Corsianos 2012, p. 157):

[1] http://ngcpconference.com/

> *When I was at a roll call, everyone laughed when the sergeant told me that I would ride along with Officer X to learn about Community Policing. 'Don't worry, we're not laughing at you, we're laughing at the Community Policing part!' Even the top guys were laughing.* (Chappell 2014, p. 19)

In US research the central problems impeding successful CP implementation were identified as policing becoming too dependent on community consent, added by the disparity of problem perception regarding their severity between police and residents (Corsianos 2012). Another issue is the aforementioned disparity in the degree of commitment to the CP concept between administrators and officers. There is the suspicion that self-interest of citizens may prevail over their community's concerns, plus residents may not want to be involved in policing activities whether person to officer or in a virtual mode of communication.

On the street cop/patrol officer level, "*lack of resources*" was quoted as the main reason why CP could not fully be implemented. Officers have to respond to emergency calls, so they cannot engage with residents. Staff shortages prevent officers from making time for CP activities. Beats are covering areas too extended to provide police officers with opportunities to get into contact with residents (Corsianos 2012, p. 156).

Interview findings indicate organizational barriers in their daily work that prevent police officers from adopting community policing as activities, among them lack of officers' autonomy, ("autonomy is necessary to develop more intimate ties with their community"; Corsianos 2012, p. 157); patrol officers are resistant to change; the traditional police cultural norms value crime fighting and law enforcement higher than community building and problem-solving; police training has not been supplied and teaching has not been adapted in regards to CP practice (ibid., p. 160). This is supplemented by a general (and well-documented) police subcultural resistance to innovation. Police agencies tend to be hierarchically organized and centralized bureaucracies which are by definition resistant to change. Police organizations are more likely to change in ways that reinforce the traditional bureaucratic structure and less likely to change in ways that challenge that structure. CP challenges bureaucratic structures because of its emphasis on decentralization (ibid.). In conclusion,

> "*past research suggests that patrol officers have more negative attitudes toward problem solving and Community Policing than those in the higher ranks... Some officers are cynical or simply laugh at the idea of Community Policing*". (Chappell 2009, o. 22)

The research study by Brainard and Edlins (2014) investigated social media usage of police departments in the ten most populated US cities. The research started with the proposition that

> "*the core of Community Policing is engagement between police and external stakeholders*' for which they listed residents, local non-profit organizations, and private actors as examples. They went on that '*social media may facilitate ..., and enhance the ability of officers, administrators and citizens to understand each other better, build social capital*' and accordingly develop a '*shared sense of purpose and shared understanding of problems and solutions*". (Brainard and Edlins 2014, p. 3)

Choosing the top ten US police departments as a sample and restricting their investigation to Facebook, Twitter and YouTube, the authors looked into the use of social media by departments, the presence of departments on social media, police–citizen interaction via social media, and whether this interaction was indeed "dialogic", not simply police departments distributing information, and also whether social media would enhance the collaboration and community engagement aspects of community policing.

The research findings (over a period of 3 months) indicate the dominant role of Facebook as the by far most popular platform. There are more resident responses to police-initiated postings than police department responses to resident communication. Only one Facebook and two Twitter threads in two cities qualify as "remotely collaborative in nature" (Brainard and Edlins 2014). Chicago, a city with a long-standing and well-researched practice of CP (Skogan 2006), had the lowest number of posts. This can serve as an indicator of the above-mentioned differences between practical CP and virtual (digital) CP communication. On the whole, residents seem more active than police. Police departments tend to end the conversation they start by not responding to residents—they drop out of the conversation, "a problem that appears in online Community Policing is to some extent police department disengagement" (Brainard and Edlins 2014, pp. 13–15). Patterns of erratic police online communication over social media lead to citizen frustration. The absence of ongoing conversation is less a technology challenge but a human resources issue. Police officers' reliable engagement is of paramount importance (DPZ 2017), that is, technology as such and by itself is not likely to achieve improved patterns of CP service (see also Chaps. 4 and 8).

Findings mostly based on research in the USA suggest a number of remedies to support the practical implementation of CP. Changes in management are indispensable, otherwise there is a lack of managerial support for CP. As "officer discretion increases as one moves down the hierarchy", police officer in-service training and recruit education have to emphasize the service function of policing. According to research, they predominantly transport an adversarial relationship between the officer and the community. Patrol officers appear to "adopt the beliefs of their immediate supervisors", thus, middle management is the transmission belt for changes in attitude towards CP as service provision.

At present, and clearly as a consequence of 9/11, large portions of US law enforcement appear to have turned 180 degrees towards a homeland security, if not a military model (Lee 2010; Turner and Fox 2017). CP was often "*trumpeted as a panacea for police departments that had lost legitimacy among the communities they served*" (Lee 2010, p. 347). With Trump as an example of hyper-masculinity of the Neanderthal type, substantial change is not very likely. Clinton and Obama have actively supported a turn from zero-tolerance policing to community policing; Donald Trump seems to be more inclined to support a return to a masculinist colonial type of law enforcement.

Police and Social Media: A Double-Edged Sword

Anywhere in Western societies, evaluation of CP performance, in-service training/police college/university education and middle management engagement emerge as indispensable fundamentals for a successful practical implementation of CP as a proactive strategy and as a—in the long run—sustainable instrument of crime reduction. In case such fundamentals are lacking, it seems doubtful that technological solutions, IT gadgets and social media presence will change the climate between police and citizens. Clearly, in order to "engineer community safety" much more is needed than a smartphone with an app.

INSPEC^{2}T research attempts have included qualitative research data (observational, in-depth interviews, etc.) to account for the gap between "philosophy" and practical implementation. "Best practice" examples have had an effect on the design of the INSPEC^{2}T IT application. However, as indicated above, the very architecture of such projects as a co-operation between IT/software experts and police research has led to a sometimes subtle but occasionally rather obvious resurgence of the "police as crime-fighting" paradigm. Early drafts/versions of application cases had robbery, theft of a notebook, later a children's red bicycle being misapprehended and finally a case of break and enter in a holiday resort to demonstrate the functionality of the digital device. It remains to be seen whether this will be balanced against the community-oriented aspects which are the core objectives of what the description of work of INSPEC^{2}T and related security research projects has promised to achieve.

Since 2010 and across German Länder and Federal policing, social media communication, although to differing degrees, is now an integral part of law enforcement PR activities. Social media officers were appointed after specific training or as lateral entries into the service. In the police departments of larger cities such officers provide social media communication on a 9-to-5 schedule; however, social media monitoring is carried out on a 24/7 basis. Accordingly, cities like Karlsruhe (pop. 300.000) have gathered 15.000 "fans" on Facebook (6000 on Twitter), and Neu-Ulm (pop. 60.000) police have 7000 Facebook and 1000 Twitter "fans" (DPZ 2017 p. 9). Social media officers list the following advantages of such tools, mostly in the area of informing the public and to correct false information and rumours, and to find witnesses after traffic accidents; for example, a police-issued post about an accident involving a child victim was shared ten thousand times and reached a total of one million users, which resulted in a witness coming forward. Other useful effects are seen in the fast and efficient ways social media help police to do the following:

- Issue warnings.
- Issue traffic/accident information.
- Get citizens' information after incidents.
- Inform about environmental dangers.

A social media officer of Berlin police is quoted on the issue of humour: "*If police are prohibited from using humour, this is bad for police and for the country*" (Süddeutsche Zeitung Magazin 2017, p. 13). She defended the use of a post where police pokes fun at gender-correct language. She goes on to say that it has been decided that "police cannot be offended" on social media. Only if offending user messages infringe upon criminal statutes, charges are laid against users from whom such posts have originated. Also, Berlin police reports such posts to the social media provider, and in the case of Twitter, the officer states that such posts (e.g. hate speech, xenophobic or racist) had been deleted.

The possibility of immediate correction of false information by users that could infuse a public panic during an attack was used in a particularly efficient manner by Munich police social media (PR) officer Marcus da Gloria Martins, a former Master student of my university (German Police University DHPOL). His interventions in the course of the mass-shooting attack in the Munich shopping mall (Olympia-Einkaufszentrum 22nd July 2016; cp. Chap. 6) rectified many of the rumours spread through social media in the aftermath of the assault that left nine persons dead and five wounded. The attack was carried out by a young man, and similar to other mass shooting incidents, it was an act of revenge against his peers. However, this shooting attack occurred during a time of Islamist terror incidents in other European cities. When, for instance, Munich users had spread the rumour that "*men with rifles*" were seen at one of Munich's busiest city squares, this and other false claims resulted in a panic that, independent of the shooting, left over 30 persons injured. Officer da Gloria Martins's calm responses helped to alleviate the mass anxiety about an allegedly IS-related act of terror and may have averted a veritable mass panic situation. For this example of professional use of social media, Mr da Gloria Martins and the Munich Police PR department got nationwide recognition.

In a difficult situation for the city, its inhabitants, and the city's police department, da Gloria Martins' social media activities were called "*the voice of reason*" in an outbreak of social media mass hysteria where most tweets misleadingly indicated Islamic terror. Mr da Gloria Martins and his colleagues won an award from Germany's Federal Public Relations Association. Although he was a modest person ("*celebrity status and being a cop are not compatible*"), da Gloria Martins gained a prominent status as a police officer and was handed around in German and international television talk shows. In an interview (DPZ 2017), he criticized Facebook for having wasted too much of his own time and said that it has mutated into a "*private high-gloss direct marketing*" device. He acknowledges the advantages of police social media engagement as "*getting access to parts of the citizenry that would otherwise never actively contact police.*" Social media involvement has generated positive responses from those citizens. Da Gloria Martins insists that the actual running of police social media communication, the editorial control and the selection of topics ought to remain exclusively under the auspices of police authorities (Interview DPZ 2017). His standpoint is pretty much in line with that of other big-city police departments (Frankfurt, Hamburg and Bonn) who tend to be critical of external, private crime/security enterprises like *Dike* (the Greek goddess of justice) and software corporations running apps for profit-oriented purposes. In Germany, fundamental scepticism against crime-watcher apps (one of them actually named "Block watcher"—a reminiscence, intended or not, to the Nazi *Blockwart* system) is shared by data protection agencies and police alike (see also Chap. 7 on digital vigilantism).

Undeniably, social media usage has negative repercussions for police work. Most prominently, large groups of sensationalist smartphone owners take camera or video footage of accident or crime scenes, and put these images on Facebook, Twitter or similar social mass media. As a direct effect, they block access to the accident location or crime scene. There have been reports that groups of this new breed of non-helping bystanders have actively abused police, because they were asked to stop the filming of victims and police. Police of a mid-size city near Dortmund in Westphalia had to publicly admonish social media users who in their avarice for sensational pictures/videos blocked access to an accident scene where the victim, a young girl, was about to die from her injuries. In other instances, social media footage posted by sensationalist bystanders was not useful to law enforcement but to the offenders because they could figure out where police personnel was located or coming in. As we know from various studies of mass-shooting assaults ("amok") potential copycat offenders are known to liberally use such media material to plan their own acts. During the aforementioned shooting in Munich and other incidents, social media officers had to warn sensationalist users that they risk prosecution if they post footage or pictures of victims or police during rescue or crime investigation activities.

In an essay published by Süddeutsche Zeitung, Austrian critic Karl-Markus Gauss (2017) raises a number of pertinent questions regarding social media bystander catastrophe sensationalism that ought to be covered more deeply by IT and sociological/media research. Owing to our brain structure's heritage, humans instinctively attempt to observe dangerous situations and catastrophic incidents. Turning one's head as a car driver to observe an accident scene on the opposite lane of the expressway or a burning car can be explained as an instinctive reaction. However, stopping one's car on the emergency lane (which may cause further collisions), climbing out, activating the cell phone's picture or preferably, video function, joining fellow digital onlookers, zooming in on injured or dead victims and afterwards immediately posting this on social media cannot reasonably be excused as behaviour led by human instincts. It is Gauss's suggestion that the expectation to cover catastrophe by social media "documentation" has become the centre, the focal concern of too many people's digital *Lebensbewältigung*, their coping with digital technology *and* real-life situations. To carry a smartphone as an extension of the body's natural senses and an artificial memory support system is perceived by some as a human right of the first order: the smartphone as a body part. Catastrophe could occur at any moment; that is why you always have your cell phone at hand.

"Being there", a film based on a novel by Jerzy Kosinski with Peter Sellers as the lead actor, tells a story of Mr Chance, the gardener, who has lived in seclusion and whose entire knowledge of the world derives from gardening and watching television (Being There 1971). When the wealthy owner of the Washington, D.C. house where Mr Chance has worked the garden dies, Chance wanders into American reality. Chance carries the TV remote control with him, and when he encounters unpleasant scenarios (e.g. street gang members trying to harass him), Chance pushes the remote buttons because he believes by using the remote, he can switch into a different channel.

Whether in public transport, at a café or a park, a children's playground or even at a jogging path by mothers or fathers pushing a stroller, our world seems populated by millions of descendants of Mr Chance, brave new world smartphone zombies, who are looking at their digital devices, pushing buttons and waiting for messages from other users—or for something sensational to occur in front of the lenses of their cell phone or pads so that they can post a picture on Twitter or a clip on Facebook or YouTube. Gauss (2017) concludes his essay with the statement that the expectation of "life" catastrophic scenes, to be "saved" on digital technology by such myriads of users, is the "true catastrophe" of our times. Sceptics will put community policing apps into the wider realm of such pathological tendencies towards sensationalism.

Domestic Violence

Some practices enabled by IT and cell phone technology can be considered as "good practices" in light of being useful for policing and prevention purposes and can be integrated in a platform and disseminated. As an example in the context of domestic violence, Berlin police co-operation with a mobile phone company has resulted in the equipment of high risk individuals with a cell phone. With this, victims have immediate access to police by punching a single key. A CP app, as it is developed and tested across the participating EU member states, bears an excellent chance for police regarding the prevention, prosecution and mitigation of high impact domestic violence. Research findings by national and international agencies consistently indicate very low reporting rates for violence against, spouses, women and other family members.[2] Reasons are, according to such studies and surveys, a lack of trust in police, negative experiences with police and other services, high threshold to report when domestic violence is witnessed (by neighbours, friends and community members; cp also Chap. 5). A user-friendly police–community app could have a positive impact on the reporting behaviour, particularly where police have insufficient community ties (e.g. minorities, migrants and refugees). In the latter parts of the population social and cultural factors are at work that decrease trust in state authorities and police, and on the other hand tolerate or even condone the subordination of females and to a higher degree than in the mainstream society accept violence in the family. The analysis and refinement of risk assessment tools can use the data from digital police–citizen communication (cp. Chaps. 10 and 12). Empirical findings can be transmitted into police training and education practices or co-operation between NGOs, activist groups or public health agencies with law enforcement on a local basis or in a wider framework. At present, the European Union is funding comparative research that investigates the possibilities to use such modern communication technologies for domestic violence victims or witnesses to better contact emergency call centres, particularly by vulnerable groups such as disabled and the old.

In this research special consideration will be given to communities of underprivileged populations and ethnic minority backgrounds for a number of reasons. Domestic violence is regularly more prevalent in these neighbourhoods. Secondly, victims with an ethnic minority background are often less likely to report to the police or seek help from outsiders, because of mistrust, prejudice or fear of deportation.

[2] see www.improdova.eu

Conclusion

Law enforcement's usage of social media and cell phone technology as reviewed and criticized in this chapter will determine whether police will continue on a trail "*to better serve and protect*" or turn into "*21st century police techno science*" (Sheptycki and Bowling, 2016, p. 9). Strittmatter (2018) describes the all-encompassing integration of such technologies into Chinese Communist Party/Ministry of Interior police control of citizens which in hindsight makes Orwell's *1984* look like a Grimm brothers' fairy tale of olden times. The predominance of both, binary IT logics and the traditional masculinity fixated "crime-fighting ideology" as exemplified in this essay in the software development for police apps, will rarely add to more democratic and better policing. It is much more likely that it will fundamentally contradict the service-oriented approaches of "better policing".

References

Being There. (1971). *Motion Picture directed by Hal Ashby with Peter Sellers.* Shirley MacLaine and Jack Warden. Retrieved from https://en.wikipedia.org/wiki/Being_There

Brainard, L., & Edlins, M. (2014). Top 10 US municipal police departments and their social media usage. *American Review of Public Administration, 61*, 1–18.

Chappell, A. T. (2009). The philosophical versus actual adoption of community policing—A case study. *Criminal Justice Review, 34*(1), 5–28.

Corsianos, M. (2012). *The complexities of police corruption-gender, identity, and misconduct.* Lanham, MD: Rowman & Littelfield.

DPZ. (2017). Polizei und die Sozialen Medien. *Die Polizei Zeitschrift Baden-Württemberg, 1*, 6–14.

Elias, N. and Scotson, J. L. (1994). The Established and the Outsiders: A Sociological Enquiry into Community Problems. London: Sage.

Gauss, K. M. (2017). Echtzeit. Süddeutsche Zeitung Nr. 59, (11.712. März), p. 5.

Keane, N. J. (2016). Police use of social media to support community engagement—Its rise in police practice in the UK. *European Police Science and Research Bulletin.* Special Conference Edition, 101–105.

Lee, J. V. (2010). Policing after 9/11: Community policing in an age of homeland security. *Police Quarterly, 13*, 347–362.

Manning, P. K. (2010). *Democratic policing in a changing world.* Boulder: Paradigm Publishers.

Sheptycki, J., & Bowling, B. (2016). Global policing and the constabulary ethic. *European Police Science and Research Bulletin.* Special Conference Edition, 9–23.

Skogan, W. G. (2006). *Police and community in Chicago: A tale of three cities* (p. 344). Oxford: Oxford University Press.

Strittmatter, K. (2018). Augen auf – Chinas KP erfüllt sich den Traum aller autoritären Herrscher: die totale Überwachung und Kontrolle des Volkes. *Süddeutsche Zeitung, 27*, 3.

Süddeutsche Zeitung Magazin. (2017). Stopp. Bitte Lesen – Ein Heft über die Polizei. Nr. 31, 4. August.

The Guardian. (2017). *Growing social media backlash among young people, survey shows*. Retrieved from https://www.theguardian.com/media/2017/oct/05/growing-social-media-backlash-among-young-people-survey-shows.

Tönnies, F. (1887). Gemeinschaft und Gesellschaft. Grundbegriffe der reinen Soziologie. Berlin. Available online: http://www.deutschestextarchiv.de/book/show/toennies_gemeinschaft_1887

Turner, F. W., & Fox, B. H. (2017). Public servants or police soldiers? An analysis of opinions on the militarization of policing from police executives, law enforcement, and members of the 114th congress U.S. house of representatives. *Police Practice and Research*.

CHAPTER 3

Cultural Intelligence and Community Policing

Anja Louis and Helen Grantham

Introduction

In recent police history one case more than any other defined and haunted the British Police force and its attitude to race. The Stephen Lawrence murder in 1993 shook the nation: A black teenager was stabbed by a gang of white youths whilst he and a friend were innocently waiting at a bus stop. There was a reluctance to charge the prime suspects despite witness statements and various leads from the community. Even after the suspects were finally charged, the Crown Prosecution Service quickly dropped the charges due to "insufficient evidence". This racially motived crime became one of the most high profile cases in UK legal history. In 1997 Her Majesty's Chief Inspector of Constabulary commissioned a review of the police service, which concluded that "racial discrimination,

A. Louis (✉)
Languages and Cultures, Sheffield Hallam University, Sheffield, UK
e-mail: a.louis@shu.ac.uk

H. Grantham
CENTRIC, Sheffield Hallam University, Sheffield, UK
e-mail: h.grantham@shu.ac.uk

B. Akhgar et al. (eds.), *Social Media Strategy in Policing*,
Security Informatics and Law Enforcement,
https://doi.org/10.1007/978-3-030-22002-0_3

both direct and indirect, and harassment are endemic within our society and the police service is no exception. [There is] a direct and vital link between internal culture in the way people are treated and external performance" (Her Majesty's Inspectorate of Constabulary 1997, p. 18). The subsequent MacPherson enquiry (Macpherson 1999) equally concluded that the Metropolitan Police (MET) was institutionally racist. It recommended the repeal of the double jeopardy rule to allow retrial upon new evidence; this came into force in the Criminal Justice Act of 2003. As a result two of the original suspects in the Lawrence murder case were retried and convicted of murder.

As the nation witnessed this landmark case and historical legal change, part of the media coverage was not only the public story of law enforcement but also the inspiring "private" story of an unyielding spirit of the Lawrence family. Their dignity and fortitude was inspiring to us all, as they never surrendered to a patronising and culturally unintelligent police force. Since then, more recent changes in migration have dictated a shift in perception of ethnic minorities and consequently have had a profound impact on the relationship between the police and multicultural citizens (Bowling and Phillips 2003).

Debates about otherness and othering (Holliday 2013)[1] have become commonplace in contemporary societies. Yet 25 years after the Lawrence murder case cultural intelligence and other intercultural models still seem to have little impact on police community work. This chapter therefore introduces three classic intercultural models (Bennett 1993; Deardorff 2006; Earley and Ang 2003), discusses its theoretical and practical implications, positions it in the broader sphere of societal impact and then applies them to both positive and negative case studies, mainly regarding Twitter use. It concludes that knowledge of, and training in, intercultural competences would improve police work, lead to more positive outreach work and reduce intercultural tensions in multicultural communities.

Intercultural Theories

It is no coincidence that the inception of intercultural theories started shortly after World War II. After two world wars the international community, under the auspices of the United Nations, was eager to promote coop-

[1] See Adrian Holliday (2013, p. 13), where he defines othering as "reducing a group of people to a negative stereotype."

eration and intercultural understanding between nations. The generation that spearheaded this change had experienced the wars and wanted to build a better world. The anthropologist Edward T. Hall, for example, one of the founding fathers of intercultural communication, had worked for the Foreign Service Institute of the US State Department in the 1950s. He had served in Europe and the Philippines; working in intercultural spaces he learned "first-hand about the details and complexities of one of the world's most significant problems: Intercultural relations" (Hall 1992, p. 76).

In what follows we give an overview of three theoretical models that are particularly suited for intercultural training, since they bridge the gap between theory and practice. Both Bennett's and Deardorff's models focus on an individual's exposure to other cultures, whereas Earley and Ang's four dimensions—whilst still focussing on the individual—measure cultural intelligence. Whilst Bennett explains the linear process we undergo when experiencing difference through exposure to other cultures (and therefore potentially enhance self-awareness), Deardorff equips us with a pragmatic developmental model of intercultural competence. Earley and Ang's contribution lies in the reconceptualisation of intercultural sensitivity or competence into the concept of cultural intelligence and its claim of measurability. Let us now look at each model in turn.

Milton Bennett's Developmental Model of Intercultural Sensitivity (1986)

One of the classic models of intercultural studies is Milton Bennett's developmental model of intercultural sensitivity (1986). As an individual is exposed to a foreign culture, ideally they work through six stages of development (see Fig. 3.1):

- ***Stage 1 'Denial of difference':*** The individual is ignorant of cultural differences and hence does not see the need to change. Benign stereotyping goes hand in hand with aggressive ignorance. At this stage the individual might say things like "as long as we speak the same language, there's no problem."
- ***Stage 2 'Defence of difference':*** Once the individual has had enough contact with another culture, they perceive difference and react defensively to it. Cognitive categories of difference are assessed negatively and cultural superiority communicated in binary opposites (them/us): "when you visit other countries you realise how much better we are."

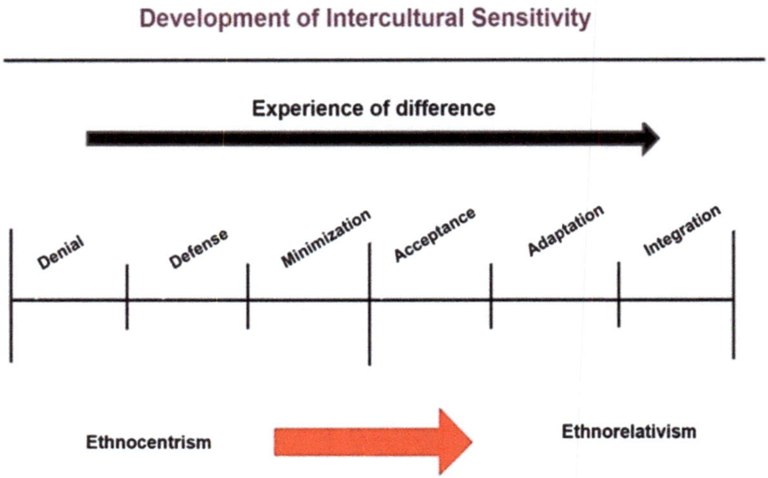

Fig. 3.1 Bennett's six stages of development (adapted from http://www.idrinstitute.org/page.asp?menu1=15)

- ***Stage 3 'Minimization of difference':*** By this stage the superiority is overcome, individuals now minimise difference and emphasise the commonality of basic values. Individuals might say: "customs differ, but as people they're really like us."

Whilst stages 1–3 are ethnocentric in their worldview, stages 4–6 are ethno-relative.

- ***Stage 4 Acceptance of difference:*** This stage is a deeper acceptance of, and respect for, cultural difference. Individuals can distinguish between beliefs, values and behaviours of diverse cultures, refrain from value judgements and analyse culture at a metalevel. At this stage the individual might say, "We are all enriched by diversity."
- ***Stage 5 Adaptation to difference:*** The individual has adapted to the other culture and can consciously shift their frame of reference in order to behave in culturally appropriate ways.
- ***Stage 6 Integration:*** The individual has completely integrated in a new culture. Their identity is not based on any one culture but comfortably moves from one cultural frame to another in a multicultural society. Here the individual says things like, "I feel most comfortable when I'm bridging the gap between the cultures I know" (Bennett 1993).

Deardorff's Model of Intercultural Competence

In contrast, Carla Deardorff's Model of Intercultural Competence is a circular model which works through the prerequisites and process of transforming individuals into interculturally competent people (see Fig. 3.2). The prerequisites are valued-based attributes such as respect,

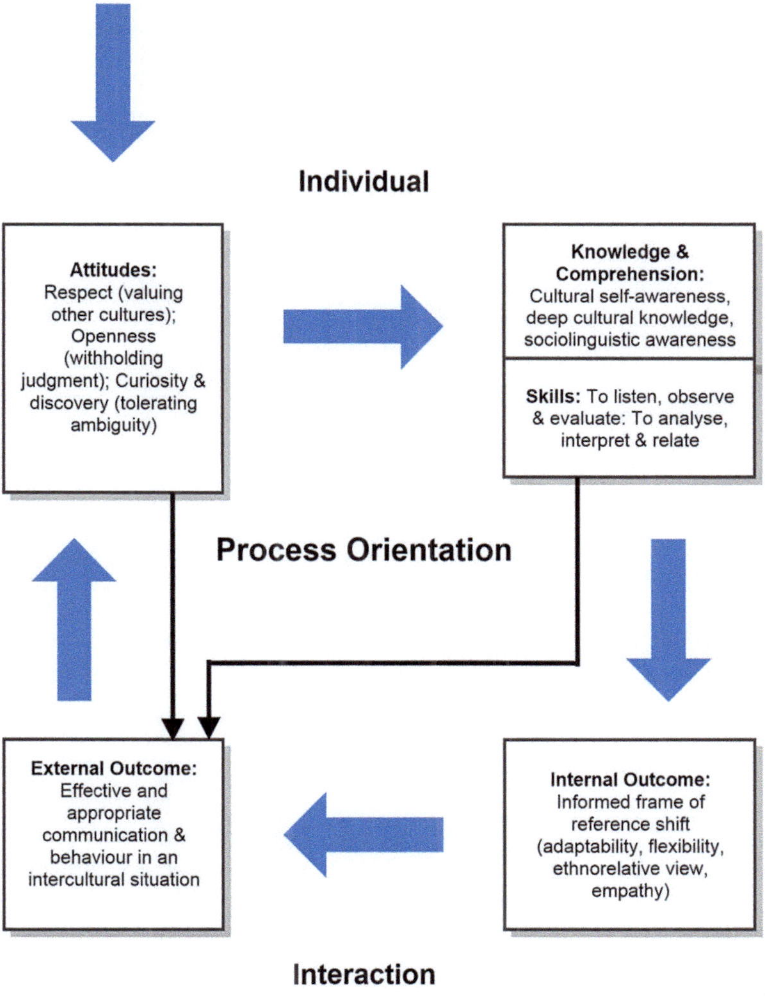

Fig. 3.2 Deardorff's Model of Intercultural Competence (adapted from Deardorff 2006)

openness (withholding judgement) which leads to non-judgemental observation, a genuine curiosity for the other cultures and, most importantly, tolerance for ambiguity. This works hand in hand with deep cultural knowledge as well as cultural self-awareness (knowing one's own biases and thought processes). The desired internal outcome is to be able to shift frames of reference and be able to view the world with somebody else's eyes. Understanding others' worldviews, empathy and adaptability are the internal outcome of this approach. This then has to be evidenced in behaviours at a communicative (verbal and non-verbal) as well as action level (external outcome; Deardorff 2006).

Cultural Intelligence

In the 1980s conventional views on intelligence and its measurement (IQ) were superseded by more encompassing conceptualisations of intelligence and its domains. "*Howard Gardner's seminal Frames of Mind: The Theory of Multiple Intelligences*" (1983) expanded the hitherto narrow focus on academic intelligence (linguistic, mathematical and visual-spatial) to a broader definition and inclusion of musical, kinaesthetic, interpersonal, intrapersonal and naturalistic intelligence. This widening of the definition was an attempt to empower individuals and create a system which does not only favour academically intelligent people. Equally, Sternberg and Detterman's "What is Intelligence?" (1986) and Sternberg's subsequent notion of "successful intelligence" (1997)[2] reconfigure conventional views on intelligence to give it real-world relevance.

Earley and Ang built on this growing agreement that intelligence is a multidimensional concept and mapped their conceptual model of Cultural Intelligence (CQ) on these ground-breaking conceptualisations of intelligence. They introduced the concept of cultural intelligence in 2003 at a time when unprecedented globalisation increased the need to resolve misinterpretations arising from cultural barriers and—from a political perspective—in the aftermath of 9/11 with its ideological clashes and cultural conflicts. Their research question was simple ("Why do some individuals function more effectively than others in culturally diverse situations?") and subsequently defined

[2] In this paper he defines successful intelligence as "the ability to achieve one's goals in life given one's sociocultural context ... through a combination of analytical, creative and practical abilities." For further reading, also see Sternberg (2005).

cultural intelligence as "the capability of an individual to function effectively in situations characterised by cultural diversity" (Earley and Ang 2003, p. 59)[3].

This conceptualisation and method of measuring cultural intelligence attests to a desire to reflect the ability to learn (intelligence as a malleable construct) and potential to succeed. Similar to an IQ that measures one's analytical and problem-solving skills, CQ measures those skills, amongst others, applied to culture. Cultural intelligence as a multifactor construct consists of four factors: *metacognitive CQ* (knowledge and control of cognition, the process the individual uses to acquire and understand knowledge); *cognitive CQ* (individual knowledge and knowledge structure); *motivational CQ* (most cognition is motivated by something and thus it focuses on magnitude and direction of energy as a locus of intelligence); and *behavioural CQ* (focuses on individual capabilities at an action level). Let us now discuss these in turn.

Metacognitive CQ

Metacognitive CQ focuses on high-order cognitive processes (thinking about thinking) and denotes somebody's conscious cultural awareness during intercultural encounters. People with a high metacognitive quotient are able to question their cultural assumptions, interrupt their thought processes and are critically aware of cultural stereotypes. A higher-level metacognitive CQ enables individuals to develop novel approaches to new cultural environments, mainly because the information-processing takes place at a deeper, self-critical, level; they thus adjust their cultural knowledge when interacting with others. A test question might be: "What do I *think* about this culture?" This factor is vital for cultural intelligence, because it challenges our assumptions, revises stereotypes of cultural norms and makes adaptation to difference and diversity easier. Ideally, in a virtuous learning circle we learn from each intercultural encounter and thus add to our databank of diversity.

Cognitive CQ

Cognitive CQ reflects the knowledge of different cultures and societies. Similar to Deardorff's Knowledge/Comprehension/Skills requisite for intercultural competence, cognitive CQ measures knowledge at a theoretical and practical level. Theoretical knowledge can be acquired through a variety of reliable information sources (books, articles, websites and the media) whilst practical knowledge can only be gained through exposure to the culture when ideally the practitioner becomes an observer or kind of

[3] For further reading, also see Ang and Dyne.

cultural anthropologist.[4] A test question might be, "What do I *know* about this culture?"

Motivational CQ
Motivational CQ refers to the source of initiative in intercultural settings (the willingness to get things done) as well as the ability to focus energy and affect to function in intercultural situations. The difference between latent capacity and developed ability can be overcome by a high motivational CQ. Based on intrinsic motivation those with a high CQ will bring higher energy levels to the completion of an intercultural task, therefore expect to successfully accomplish the task and appreciate the value of the accomplishment.[5]

Behavioural CQ
Behavioural CQ defines the cultural appropriateness of verbal and non-verbal acts. When we interact with people from diverse cultures our behaviour has to be mindful of the range of acceptable behaviours and the interpretation that different cultures might attribute to styles of interaction. In intercultural communication, for example, Edward Hall's concepts of high and low context cultures are helpful knowledge when communicating across cultures (cognitive CQ; Hall 1976). Cultures that rely on explicit, direct communication (you need little information from the context to make sense of the statement) are low context cultures (e.g. Germany, Scandinavian countries, the USA), whilst in high context cultures (e.g. China, Korea, Japan, Arab nations) the communicative style emphasises non-verbal messages. Since communication is indirect and/or unspoken, you need to know the cultural context to make sense of the situation. People with high behavioural CQ can translate their cognitive knowledge (e.g. Hall's conceptualization) into interculturally appropriate behaviour. Test questions here might be, "How do I have to behave in a low context culture for successful communication to take place? What exchange of verbal and non-verbal messages has to take place between the individual and the other culture to develop mutual respect and trust?" Whilst all factors of this multidimensional construct are important, the behavioural CQ is the most visible manifestation of cultural intelligence to

[4] For interesting case studies of cultural practices, see Holliday (2013)

[5] For further details of the value-expectancy theory of motivation, see Eccles and Wigfield (2002)

the outside world. The adage of "actions speak louder than words" is particularly pertinent in intercultural contexts.

What is the difference between emotional and cultural intelligence? There is, of course, overlap in the sense that some dimensions of emotional and cultural intelligence are the same, most obviously interpersonal (understand the intentions and motivations of other people) and intrapersonal intelligence (understand oneself) (Gardner 1983). However, somebody with high emotional intelligence might only be able to interact successfully in their own culture with culturally similar individuals. In contrast, a person with high cultural intelligence "can somehow tease out of a person's or group's behaviour those features that would be true of all people and all groups, those peculiar to this person or this group, and those that are neither universal nor idiosyncratic. The vast realm that lies between those two poles is culture" (Earley and Mosakowski 2004, pp. 139–144).

Since its application lies in cross-cultural situations, to a certain extent cultural intelligence is a culture-free construct; its metacognitive factor gives us the necessary ability to shift cultural frames easily.[6] We are all influenced by discourses of culture, stories we tell each other to explain who we are. This is usually based on reliable and unreliable information, often propagated by the media, and once we invest emotionally in particular stories that narrative will then take precedence in our minds. Stereotypes as short-hand categorisations of, and assumptions about, otherness abound.

This works in both directions, of course, ethnic minorities might have a stereotypical view of the police as much as police officers on the beat regularly thin-slice individuals. A culturally intelligent person suspends judgement and thinks before they act (another overlap with emotional intelligence). Cultural intelligence forces you to switch off "cultural cruise control" (Thomas and Inkson 2009), a form of automated behaviour based on (mono)-cultural assumptions that then clash with norms from other cultures. Thomas and Inkson argue that we have to move from this state of mindlessness (not critically thinking about our cultural assumptions) to mindful non-judgement.

Paying attention to the intercultural encounter whilst having a beginner's mind of curiosity can be instrumental in an enhanced CQ. Mindfulness then is facilitating a shift from theoretical knowledge and self-awareness to

[6] A full discussion of the correlation between emotional and cultural intelligences is beyond the scope if this chapter. For an excellent discussion see Moon (2010).

proficient intercultural practice.[7] Ideally, as an observer and cultural anthropologist we observe non-judgementally how people interact in a different culture and how this is different from our own culture. These observations take time and require us to pay focussed attention to our surroundings with an open mind. Once observed behaviour falls into patterns, we can predict cultural differences whilst not falling into the traps of selective perception (we only see what we want to see) or a combination of attribution, social categorisation and stereotyping: "He hits his wife because he is Eastern European; they are overweight because they're Indian; she isn't trustworthy because she's black" (Thomas and Inkson 2009).

Cultural Intelligence in Community Policing

Community Policing promises to change radically the relationship between the police and the public, address underlying community problems, and improve the living conditions in neighbourhoods (Eck and Rosenbaum 1994). There has been much academic discussion about the success of applying Community Policing and how police departments and their officers worldwide can use this innovative method of policing to connect with their community on a deeper level. Trojanowicz and Bucqueroux (1990) affirm this in their discussion on Community Policing, stating that Community Policing changes the fundamental nature of the relationship between people and their police to one of mutual respect and trust. Furthermore, by building trust between the community and law enforcement it allows officers to develop knowledge of the community and resident activity and can thus provide vital intelligence relating to serious crime and potential terrorist acts (Docobe 2005; see also Chaps. 6 and 8).

Recent shifts in migration and demographic changes increase the diversity in our modern societies and require police forces to seriously consider their role in moderating complex intergroup relationships in the communities they serve (Degeneste and Sullivan 1997). London, for example, has the fastest changing and growing population in the UK with over 100,000 people joining the current 8.6 million Londoners who already inhabit the city. This growth reflects an ever greater diversity of nationalities, language, religion, background, experience and beliefs. Hence police forces

[7] A full discussion of mindfulness in intercultural communication is beyond the scope of this chapter. For further details, see Chang (2013), Gudykunst (2005), Langer (2003) and Shafir (2009).

in multicultural societies need to consider the challenges of policing racially and culturally diverse communities in order to achieve a positive outcome (King 2016).

However, judging from the dearth of writing on this subject there seems to be a lack of consideration for cultural intelligence. Neither in academic nor in practitioners' journals did we find evidence of any meaningful debate. Whilst police forces themselves have been given the reasoning for implementing community policing there is not much evidence to suggest that they hold the appropriate level of cultural intelligence to implement it successfully. Cultural intelligence is a model that police forces can utilise to understand the underlying cultural needs of the communities in which they are working. This would give them the skill set to understand cultures different to that of their own and use that knowledge to communicate effectively with their community, build their trust and understand why individuals from different cultures behave in a certain way.

Without this invaluable knowledge there is a danger of irreversible damage. Building relationships amongst communities who have the same cultural background is less problematic, but when the police and the community have differing cultural backgrounds that are not reflected in their policing methods it can lead to feelings of mistrust, thus ultimately leading to the breakdown of the key relationships between police forces and the communities they serve. To a certain extent, it is up to the police officers to make themselves worthy of this trust. In what follows we will discuss the Stephen Lawrence case as a historical example, then discuss four recent case studies of community engagement and analyse them according to the intercultural models of Bennett, Deardorff and Earley and Ang.

The Stephen Lawrence Case

The Macpherson inquiry was instigated due to allegations that the police did not act appropriately at the scene of Stephen Lawrence's murder and during the subsequent investigation. The summary of Macpherson's report (Macpherson 1999) discusses how a combination of investigative failures, fundamental errors and, ultimately, institutional racism undoubtedly weakened the first investigation into Stephen's death. Some examples of irresponsibility summarised in the report are as follows:

- **First Aid:** No police officer did anything by way of First Aid, apart from testing to see whether Stephen Lawrence was still breathing and whether his pulse was beating.

- **Family Liaison and Victim Support:** From the first contact with police officers at the hospital, and thereafter, Mr and Mrs Lawrence were treated insensitively, with a patronising attitude and lack of sympathy. Most importantly, they were never given information about the investigation to which they were entitled.
- **Surveillance:** The surveillance operation was ill-planned, badly carried out, and inadequately documented.
- **Identity Parades:** The identification parades were poorly planned. There were clear breaches of the Codes of Practice governing identity parades.
- **Searches:** When the arrests were made, it is plain that the searches of all the suspects' premises were inadequate.
- **Racism:** Inspector Groves's (the only senior officer present before the ambulance came) insensitive and racist behaviour at the scene. He assumed that there had been a fight. He wholly failed to assess Duwayne Brooks (Stephen's friend who was also injured) as a primary victim. He thus failed to take advantage of the help which Mr Brooks could have given.
- **The use of inappropriate and offensive language:** Racism awareness training was almost non-existent at every level.
- At least five officers simply refused to accept that this was a purely racist murder.

Several senior police officers later admitted the problem of racism within the force. The National Black Police Association (NBPA), the biggest group representing minority officers in the force, also found overwhelming evidence that racism was part of the police culture during the 1990s, and in turn this shaped police practices deployed on the street (Bowling 2018). Research has shown time and again that in police-initiated encounters black and ethnic minorities are disproportionally often stopped by the police (Bowling and Phillips 2003). The purpose of stop-and-search is predominantly used for crime prevention or detection and as such an investigative power (Lustgarten 2002). However, it is also frequently used in order to break up groups of young people and for social control (FitzGerald 1999; Waddington et al. 2002), despite the fact that Home Office research demonstrated that the strategy has a little impact on crime deterrence (Bowling and Foster 2002; Miller et al. 2000). The Macpherson report also concluded that racial stereotyping and prejudices were contributing to the ways in that routine police duties were being carried out. Failing to respond appropriately to racist attacks as well as over-using their stop-and-search powers on the street resulted in a catastrophic loss of confidence and trust in the police from the black community (Bowling 2018).

The Lawrence case opened the eyes of the black community, making them much more aware of the dangers they faced, thus converting the police into more of an enemy than a protector. Too many years of mistreatment and hostility from the police towards the black community have passed. 25 years on from Steven's death and the problems of racism and inequality can still be found behind the succession of violence that is currently displayed in the horrific knife crimes taking place across London and other cities across the UK.

The police's heavy-handed, and controversial, stop-and-search methods are particularly problematic when policing minority ethnic communities. Bowling and Phillips (2003) quote the former MP for Haringey Bernie Grant who stated that "For young black people in particular, the humiliating experience of being repeatedly stopped and searched is a fact of life, in some parts of London at least. It is hardly surprising that those on the receiving end should develop hostile attitudes towards the police" (Bowling and Phillips 2003, p. 7).

On the tenth anniversary of the Stephen Lawrence Inquiry the NBPA (2009) published a report which concluded that the perceived watershed moment and subsequently predicted paradigm shift in police attitudes towards ethnic minorities had not materialised (NBPA report 2009). On the eve of the twentieth anniversary of Stephen Lawrence's death Scotland Yard's black and Asian police officers made a dramatic intervention when they declared that the Metropolitan Police are still institutionally racist, stating that "Despite the training and community initiatives put in place over the past two decades, Scotland Yard has failed to tackle the mind-set at the heart of failures over Lawrence" (Muir 2013).

Worse still, even in the post-Macpherson era, minority ethnic police officers have been subjected to abuse by colleagues in order to test their professional commitment to the job (Cashmore 2001). The NBPA, which has 500 members in London, points out that the MET has failed to recruit enough minorities to reflect the capital's population: "The 2011 census indicated that over 40% of Londoners were from BME (Black and Minority

Ethnicity) backgrounds, whilst only 10% of MET police officers were from BME communities" (Muir 2013).

According to NBPA president Tola Munro this has decreased even further to 6.3% in 2018 (BBC News 2018). These officers disproportionately hold lower-ranked positions, face significantly slower rates of career progression and are over-represented in disciplinary actions, in comparison to their white counterparts. Government figures equally show how much more progress needs to be made:

- There is no police force in England and Wales that has a BME representation that matches its local demographic.
- According to self-declared data submitted by forces to the Home Office, four forces have no Black/Black British officers.
- There are no BME chief constables.[8]

This current position of police recruitment is unsustainable, as it severely impacts on police legitimacy and more importantly erodes trust and confidence in BME communities. Furthermore, the police now need to work towards diminishing these deeply entrenched feelings by giving fair treatment to all members of the community, and by demonstrating that it aims to assist and serve everybody, not through failed archaic enforcement practices they used in previous generations, but through scrupulous efforts to problem-solve, build trust and impose fairness as a matter of course (Degeneste and Sullivan 1997).

A similar but more nuanced approach was suggested in a more recent extract from The Metropolitan Police Service (King 2016):

> *Policing is not for the faint hearted, but it needs more than determination, it needs knowledge, understanding, strong personal values, highly developed skills and perhaps most importantly the ability to cross boundaries. Our frontline staff and leaders deal with immensely challenging incidents 24 hours a day, every day of the year and need the Cultural Intelligence both to resolve them safely and to develop longer term solutions to the issues that generate police demand.*

[8] Diversity profiles published to encourage police forces to become more representative—https://www.gov.uk/government/news/diversity-profiles-published-to-encourage-police-forces-to-become-more-representative

In what follows we discuss both positive and negative case studies. Due to the scarcity of academic literature, these have been taken from media and online sources. These examples illustrate the relevance, appropriateness and urgency of applying the above intercultural models in a modern police force.

Polish Tweets

It is important to recognise good practice so that this can then be implemented by other police departments. If executed in the right manner, Twitter and other social media platforms can help to foster excellent relationships between the police and members of the public (cp. Chap. 4). An example of a Police force aiming to become more culturally intelligent was first reported in 2015. The London *Evening Standard* newspaper reported how London's Metropolitan Police were launching a Twitter account that would only tweet messages and appeals in Polish, specifically for Polish people living in London (Polish was found to be the second most spoken language in the UK in the 2011 census). The project, the first of its kind, was being trialled in Ealing and Hounslow where there is a large concentration of Polish communities (Rucki 2015).

The initiative was spearheaded by PC Magdalena Rosiak who explains that "Strong communication is vital for effective policing, particularly when assisting victims of crime or piecing together information" (Rucki 2015). It seems obvious that whilst tweeting important information in English, they would also tweet in different languages, yet this practice appears to be quite an unusual one. By communicating in Polish, the Police are promoting relations between themselves and the Polish community, consequently increasing social inclusion amongst a community who may or may not speak very good English.

PC Rosiak is clearly in Bennett's stage 5 (adaptation) as she tries to adapt to the reality of a multicultural community. In stage 5 the individual consciously shifts frames of reference to behave in culturally appropriate ways. In this example language skills are used to reach out to the largest ethnic group in London. Furthermore, this is another example of Deardorff's *sine qua non* of knowledge of a community and language skills to improve multicultural relations.

Rosiak displays high cultural intelligence of all four dimensions; her behaviour of using the Polish language as a means to communicate and bridge the gap between communities demonstrates she can convert her motivation into action. It is noteworthy that her motivation to reach out to Polish people is combined with pragmatism to enhance effective policing. Strained relationships between the Police and minority groups are not unique to Britain. Research from Bowling (2003) compared policing in the USA, Australia and South Africa with that in Britain and found similar concerns. Let us now turn to three examples from the USA in order to compare practices.

Boston Black History Month

A 2017 report commissioned by the Spotlight Team of the (Boston Globe 2017) found that whilst the city had made progress it also found signs of "striking inequality between black and whites" as well as signs of discrimination in a city that has only around 50% of white population (CENSUS 2010). A recent poll commissioned by the Boston Globe found that 54% of black people across the country rated Boston as "unwelcoming to people of colour", far more than other cities such as New York, Atlanta and Chicago.

Other findings in the report stated that black families living in Boston are much poorer than white ones and are less likely to own their own homes or businesses. Segregations also exist in healthcare, jobs and college enrolments. Against the backdrop of the above racial divisions the Boston Police Department committed a major error of judgement in its public relations work on Twitter during Black History month (February 2018), during which the Boston Police Department tweeted a Black History Month tribute post that honoured a white man. Capelouto (2018) of the Boston Globe described how the tweet celebrated the accomplishments of former NBA (National Basketball Association) team Boston Celtics' Coach and president Red Auerbach. The tweet applauded Mr Auerbach for being the first NBA coach to bring in a black player, field an all African-American starting line-up and hire the league's first African-American head coach.

Needless to say, the response to the tweet was not appreciated by the residents of Boston and the Police Department faced heavy criticism, with many Twitter users calling for an apology saying that the tweet was "tone deaf", "clueless" and had completely missed the point of Black History Month. Boston Police Department quickly retracted the tweet stating that their intentions "were never to offend."

As a means of damage control the department's account later posted another tweet honouring Bill Russell, the first African-American head coach in the NBA. The Huffington Post (2018) quoted the Mayor of the City, Mayor Walsh, calling the tweet "completely inappropriate and a gross misrepresentation of how we are honouring Black History Month in Boston". Many people were still not convinced that the apology tweet conveyed that the department understood why the initial tweet was so offensive.

The tweet can be interpreted in a variety of ways. At its best, it was an innocent mistake of benign ignorance of what Black History Month is and who should be celebrated (Bennett's stage 1). At worst, the inference was that it needs a white guy to give black people a chance in life as a kind of affirmative action measure (Bennett's stage 2) and hence could be seen as patronising. Auerbach might have chosen black players simply because they were the best players and would assure his professional success as a coach, in which case it was celebrating his success rather than the black players'.

On the other hand, others might argue that the tweet was celebrating a white coach who is enlightened enough to pick black players. Whatever it may be, the question of how they missed the mark so badly with such a culturally unintelligent statement still remains. Whilst the police tried to engage positively with Black History Month, this had not been thought through fully—a good example of how a relatively small act with a low level of cultural intelligence can have a considerable impact. Instead of improving race relations it only made matters worse, reducing levels of communication, respect and trust between the police and the community it serves. Arguably if the Boston Police Department had a higher level of cultural intelligence this might not have happened.

Both Deardorff's model of intercultural competence and Earley and Ang's cultural intelligence dimensions would have been helpful in this situation and probably avoided the diplomatic crisis. Deardorff clearly states that two prerequisites are essential: a non-judgemental and open-minded attitude and a genuine curiosity for the other cultures. This works hand in hand with deep cultural knowledge as well as cultural self-awareness (knowing one's own biases and thought processes). Equally, metacognitive (thought processes) and cognitive (knowledge) cultural intelligence would have questioned cultural assumptions and interrupt thought processes. A useful test question might have been this: "Is a white person the right person for Black History Month?"

Iowa Police Department: Twitter as Another Tool on the Belt

As reported in the Iowa State Daily student newspaper in 2017, the Iowa police department boasts a 17,000 strong following on Twitter that is administered by their community outreach officer, Anthony Greitar. Gehr (2017) reports Greitar felt that Twitter is "another tool on his belt" so that the department becomes part of their community. The Twitter account tweets about police information and crime by mixing it with humour and light heartedness which appeals greatly to the audience of young college students who enjoy his entertaining and witty style of conveying police information.

The article also reports how this rapport with their audience has made members of their community very attentive, which in the past has been extremely helpful with residents being quick to react to real threats or emergencies. The cultural knowledge and intelligence that is being deployed by the Iowa Police has influenced the kind of language, tone and amusing manner they are using to communicate with their audience.

Humour is a notoriously delicate instrument in intercultural communication, since cultural assumptions, knowledge and sensitivities need careful consideration. However, the Iowa Police Department have a vast number of Twitter followers who are college students living in the area. They have established that students appreciate their humour and all of this has contributed to the relationship they have created, resulting in increased trust, communication and cohesion amongst the community. Their Twitter success proves that humour can be instrumental in reaching across intercultural boundaries.

Walk in Someone Else's Shoes: Houston Police Department

An excellent example of the support for cultural intelligence training comes from Houston Police Department. Houston's Police department is a majority–minority department, with 53% of its staff identifying as minority. The article "Take a Walk in Someone Else's Shoes: A Cultural Intelligence Training Experience for Law Enforcement" (2017) discusses how Houston, the fourth largest city in the USA with a strong 2.3 million population, recognises how its city has evolved and grown over the past 40 years. With its constantly changing demographics presenting challenges as well as opportunities for the city's public safety agencies, the police department realised that "it was not immune to cultural misunderstandings" (Haywood 2017) and decided to invest in a different type of diversity training drafting in external trainers.

The trainers developed and executed experimental and scenario-based training that was customised specifically for the police department. With the primary goal of highlighting officers' connections to the community through role play, officers were given different roles to portray and act out. Furthermore, arguably a vital component was the history lessons that were also provided in each session. Through customised videos officers learn historical information about diverse areas of Houston, the different classes of people and the transitions of each neighbourhood in the city. They also show demographic shifts so they can learn about the make-up of the city. According to the training provider one of the critical aims was to "create this idea that there has to be an 'us', it can't be an 'us' versus 'them'" (Haywood 2017). This clearly refers to Bennett's stages 5 and 6, in which the individual adapts to, and integrates in, a different culture. It also attests to Deardorff's prerequisite of a non-judgemental attitude.

This cultural intelligence training goes beyond the average diversity training offered to most organisations. By combining the two components, role play and historical information, the police officers can understand and become more deeply affiliated with the city and the people they serve and protect. In turn this will enable them to become more knowledgeable creating a more proactive mind-set in the officers as well as creating a shift from reactive to proactive law enforcement that allows them to identify more easily solutions to problems that exist within their communities. Role play exercises are an effective way of shifting frames of perception and reference. It allows participants literally to walk in somebody else's shoes. As a result, individuals can be encouraged to move from an ethnocentric (Bennett stages 1–3) to an ethno-relative position (Bennett stages 4–6). History lessons are also a good example of Deardorff's knowledge/skills prerequisite for positive intercultural encounters as well as attesting to cognitive CI.

Conclusion

It is no coincidence that intercultural theories and models were developed after 1945 and their necessity was heightened again after 9/11. Learning from the Lawrence murder case also seems to suggest that changes in the police force did not go deep enough. Diversity figures of the police workforce did not improve considerably in the 25 years since Stephen

Lawrence's murder. Even if they did, there is doubt as to how much of a difference that would make.

Community Policing is not a particularly new strategy, it has been practiced for over 30 years as a tool to engage with the community. As we have seen in the case studies, the relatively new engagement tool of Social Media can help or hinder this process. When used effectively, it provides an opportunity to develop shared respect whilst learning to trust each other. This in turn might enable police officers and their community to work together in harmony, solving problems in the community. However, policymakers, police leaders and indeed the public should also have realistic expectations about the results that Community Policing can deliver. Community policing cannot function as a standalone tool to prevent serious crime and terrorism but can be effective in targeting low-level crime and disorder (Organization for Security and Co-operation in Europe 2014).

Diversity quotas and training have not provided sufficient progress. There is recognition in some police forces that cultural intelligence is a key success factor to the effectiveness of Community Policing. Looking at the business world with its large corporations and multinational businesses, we find a plethora of literature on the subject and a variety of intercultural training options for its workforce,[9] but even the business world falls short of meaningful and deep knowledge of interculturality (Tipton 2008). Fortunately, the stakes are low; the bottom line is loss of revenue. With police work the stakes are much higher, with potential loss of life, and therefore the implementation of cultural intelligence training programmes is a moral imperative.

The example of the Houston police force is noteworthy for a number of reasons. Their recruitment strategy has resulted in a majority-minority police force which mirrors the general population. This can then take into account the knowledge and skills to establish relationships with the community based on an enhanced credibility. Furthermore, the police leaders recognised the need for cultural intelligence training with paradigm-shifting approaches such as role play and history lessons. In order to go from role play to real play we suggest involving community members in the design and delivery of training programmes.

This simulation can nurture a sense of shared undertaking between the public and the police, allowing both to see issues and challenges from the

[9] See, for example, Barak (2017), Browaeys and Price (2011), Kandola (2009), Thomas and Inkson (2009)

other's perspective, whilst increasing transparency. Police departments need to fully embrace cultural intelligence and integrate it into their work culture as the new way of doing things, utilising it as a vital skill and an important aptitude for connecting with the community. As demonstrated in the Polish tweets, part of the connectivity might include the additional skills of speaking community languages. Our case studies also clearly suggest that social media skills are only ever as effective as somebody's CQ. Twitter is not the panacea for community policing, unless a Houston-style transformation has taken place in the officers' behaviour.

We suggest that Bennett's, Deardorff's and Earley and Ang's models are the most useful tools in training situations. Bennett's model of intercultural sensitivity is an easy tool for self-awareness purposes, since the trainee can easily relate to, and identify, different stages in their own intercultural journey. The shift from ethnocentric to ethno-relative worldviews is also mapped out and easily identifiable in practice. Deardorff complements Bennett, as she sheds light on the question of prerequisites and processes of personal changes in intercultural development. Her focus on internal and external outcomes forces the individual to self-reflect as well as ask for, and expect, feedback from the external environment—a vital element in the learning process.

Finally, Earley and Ang add complexity to the conceptualisation of intercultural models. Their model teases out the importance of a four-dimensional approach and they persuasively argue that only if these four dimensions (metacognitive, cognitive, motivational and behavioural) work together in harmony, can we truly speak of cultural intelligence as a form of intercultural practice. Furthermore, measuring the capabilities functions as another self-awareness tool and shortcut to identify individual weaknesses. Creating a holistic approach ranging from the diversification of recruitment, initial training and ongoing continued professional development in cultural intelligence, this will then provide community police officers with the necessary skills and knowledge to undertake their expected role.

References

Ang, S., & Van Dyne, L. (2008). Conceptualization of cultural intelligence: Definition, distinctiveness, and Nomological network. In S. Ang & L. van Dyne (Eds.), *Handbook of cultural intelligence: Theory, measurement, and applications*. London: M.E. Sharpe.

Barak, M. E. M. (2017). *Managing diversity: Toward a globally inclusive workplace*. London: Sage.

BBC News. (2018). *Black police leader says some forces 'Still institutionally racist'*. Retrieved from http://www.bbc.co.uk/news/uk-england-42702432

Bennett, M. (1993). Towards a developmental model of intercultural sensitivity. In R. M. Paige (Ed.), *Education for the intercultural experience*. Yarmouth: Intercultural Press.

Bowling, B. (2018). Stephen Lawrence: His death changed British law forever but trust in police has yet to recover. *The Conversation*. Retrieved from https://theconversation.com/stephen-lawrence-his-death-changed-british-law-forever-but-trust-in-police-has-yet-to-recover-95091

Bowling, B., & Foster, J. (2002). Policing and the police. In M. Maguire et al. (Eds.), *The Oxford handbook of criminology* (3rd ed., pp. 980–1033). Oxford: Oxford University Press.

Bowling, B., & Phillips, C. (2003). Policing ethnic minority communities. In T. Newburn (Ed.), *Handbook of policing*. Cullompton: Willan Publishing.

Browaeys, M.-J., & Price, R. (2011). *Understanding cross-cultural management* (2nd ed.). Harlow: Pearson.

Capelouto, J. D. (2018). Boston police apologize after tweeting a black history month tribute to a white man. *The Boston Globe*. Retrieved from https://www.bostonglobe.com/metro/2018/02/11/boston-police-tweeted-black-history-month-tribute-white-man-and-twitter-was-not-happy/qsC9EsYmRxK-pNgH96R0QmL/story.html

Cashmore, E. (2001) The experiences of ethnic minority police officers in Britain: under-recruitment and racial profiling in a performance culture. Ethnic and Racial Studies. 642–659.

CENSUS. (2010) United States. Retrieved from https://www.census.gov/quickfacts/fact/table/bostoncitymassachusetts/PST045216

Chang, Y. (2013). Never mind? Mindfulness and competent intercultural interaction. *Journal of Intercultural Communication, 32*, 1–17.

Deardorff, D. (2006). The identification and assessment of intercultural competence as a student outcome of internationalization at institutions of higher education in the United States. *Journal of Studies in International Education, 10*(3), 241–266.

Degeneste, H. I., Sullivan, J. P. (1997). *Policing a multicultural community, a police executive research forum publication*. Retrieved from http://www.academia.edu/927369/Policing_a_multicultural_community

Docobe, J. (2005). Community policing as the primary prevention strategy for homeland security at the local law enforcement level. Retrieved from https://www.hsaj.org/articles/183

Earley, P. C., & Mosakowski, E. (2004). Cultural Intelligence. *Harvard Business Review, 82*(10), 139–144.

Earley, P. C., & Ang, S. (2003). *Cultural intelligence: Individual interactions across cultures*. Stanford: Stanford University Press.

Eccles and Wigfield. (2002). Motivational beliefs, values and goals. *Annual Review of Psychology, 53*, 109–132.

Eck, J., & Rosenbaum, D. (1994). The new police order: Effectiveness, equity, and efficiency in community policing. *The challenge of community policing: Testing the promises.*

FitzGerald, M. (1999). *Searches in London under Section 1 of the Police and Criminal Evidence Act.* London: Metropolitan Police.

Gardner, H. (2011 [1983]). *Frames of Mind. The Theory of Multiple Intelligences.* New York: Basic Books.

Gehr, D. (2017). *Iowa State Daily.* Iowa state police department uses twitter like another tool on its belt. Retrieved from http://www.iowastatedaily.com/news/politics_and_administration/article_5b49f04c-14b2-11e7-84a7-d32f44459577.html

Gudykunst, W. B. (Ed.). (2005). *Theorizing about intercultural communication.* London: Sage.

Hall, E. T. (1992). *An anthropology of everyday life.* New York: Doubleday/Anchor Books.

Haywood, N. (2017). Take a walk in someone Else's shoes: A cultural intelligence training experience for law enforcement. Retrieved from https://cops.usdoj.gov/html/dispatch/08-2017/training_experience_for_le.html

Her Majesty's Inspectorate of Constabulary. (1997). *Winning the race: Policing plural communities. HMIC thematic inspection report on police community and race relations 1996/7.* London: Home Office.

Holliday, A. (2013). *Understanding intercultural communication: Negotiating a grammar of culture.* Abingdon: Routledge.

Kandola, B. (2009). *The value of difference: Eliminating bias in organisations.* Oxford: Pearn Kandola.

King, H. (2016). *Common purpose blog.* Why does the met need cultural intelligence? Retrieved from http://commonpurpose.org/blog/archive/why-does-the-met-police-need-cultural-intelligence/

Langer, E. J. (2003). *Among cultures: The challenge of communication.* Belmont: Wadsworth.

Lustgarten, L. (2002). *The future of stop and search.* 603-618.

Macpherson, Sir William (1999). The Stephen Lawrence inquiry. Retrieved from https://www.gov.uk/government/publications/the-stephen-lawrence-inquiry

Miller, J., Bland, N., & Quinton, P. (2000). *The impact of stops and searches on crime and the community* (Police research series paper 127). London: Home Office.

Moon, T. (2010). Emotional intelligence correlates of the four-factor model of cultural intelligence. *Journal of Managerial Psychology, 25*(8), 876–898.

Muir, H. (2013). Metropolitan police still institutionally racist, say black and Asian officers. *The Guardian*. Retrieved from https://www.theguardian.com/uk/2013/apr/21/metropolitan-police-institutionally-racist-black
NBPA report '10yr anniversary of the Stephen Lawrence Inquiry Report'. (2009). Retrieved from http://www.nbpa.co.uk/wp-content/uploads/2011/08/Stephen-Lawrence-10-yr-anniversary-report-2009.pdf
Organization for Security and Co-operation in Europe. (2014). Preventing terrorism and countering violent extremism and radicalization that Lead to terrorism: A community-policing approach. Retrieved from https://www.osce.org/atu/111438.
Rucki, A. (2015). *Evening Standard*. Metropolitan Police launch Polish Twitter account. Retrieved from https://www.standard.co.uk/news/london/metropolitan-police-launch-polish-twitter-account-10022460.html
Shafir, R. Z. (2009). Mindful listening. In J. Stewart (Ed.), *Bridges not walls: A book about interpersonal communication*. New York: McGraw-Hill.
Sternberg, R. (1997). *Successful intelligence: How practical and creative intelligence determine success in life*. New York: Penguin Putnam. See also Sternberg (2005) The theory of successful intelligence. Inter-American Journal of Psychology, 39(2), 189–202.
Sternberg, R. J. (2005). The theory of successful intelligence. *Inter-American Journal of Psychology, 39*(2), 189–202.
Sternberg, R. J., & Detterman, D. K. (Eds.). (1986). *What is intelligence?* Norwood: Ablex.
The Boston Globe. (2017). Boston. Racism. Image. Reality. Retrieved from https://apps.bostonglobe.com/spotlight/boston-racism-image-reality/series/image/
Thomas, D. C., & Inkson, K. (2009). *Cultural intelligence: Living and working globally* (2nd ed.). San Francisco: Berrett-Koehler.
Tipton, F. B. (2008). Thumbs-up is a rude gesture in Australia. *Critical Perspectives on International Business, 4*(1), 7–24.
Trojanowicz, R., & Bucqueroux, B. (1990). *Community policing and the challenge of Diversity* (p. 14). Rockville: National Center for Community Policing.
Waddington, D., Stenson, K., & Don, D. (2002). Disproportionality in police stop and search in Reading and Slough. Summary report for Thames Valley police, unpublished.

PART II

Concepts and Frameworks

CHAPTER 4

How to Make Community-Oriented Policing Customer Oriented: A Service Design Concept for Policing in Social Media

Olavi Kujanpää, Kari Pylväs, Pirjo Jukarainen, Jarmo Houtsonen, and Jari Taponen

Introduction

Social media strategy—like any strategy—for safety and security organizations should be based on an up-to-date situational awareness: knowledge of the various security needs of citizens and communities. Using the concept of Pawson and Tilley (1997) it is important to evaluate the *underlying generative mechanisms* behind the security problems, like low levels of trust or lack of intercultural skills within the police. It is essential to ask, "What type of social media engagement works, with whom, and

O. Kujanpää · K. Pylväs · P. Jukarainen · J. Houtsonen (✉)
Police University College, Tampere, Finland
e-mail: jarmo.houtsonen@polamk.fi

J. Taponen
Helsinki Police Department, Helsinki, Finland

B. Akhgar et al. (eds.), *Social Media Strategy in Policing*,
Security Informatics and Law Enforcement,
https://doi.org/10.1007/978-3-030-22002-0_4

to what extent?" Situational awareness thus includes also a critical analysis of the available policing resources and capabilities and areas in need of improvement.

The former can be seen as a *rational* approach to strategy. It emphasizes the effective alignment of goals, targets and means and suggests that activities are organized and led on the basis of a valid situation picture. A rational approach assumes that the operational environment can be understood by a systematic analysis, organizations' destinies controlled and changed by rational management, and the members of an organization able and motivated to learn new practices (Favoreu et al. 2016).

Many guidelines for conducting strategic planning and analysis resonate with a classic SWOT analysis' focus on the organizational strengths and weaknesses in relation to environmental opportunities and threats. Strategy is combining internal resources and competencies together with external opportunities and risks. Strategy means a link between organization's overall goals and daily operations supported by action plans and management systems.

Yet strategy for the presence in social media should not be police-centric but based on the idea of creating and sustaining strategic partnerships. The old police-led security architecture seems to be gradually supplemented by a mosaic of security and safety service providers and collaborative arrangements, networks and partnerships (e.g. Hoogenboom 2010). Such new forms of networks, nodal governance and pluralistic systems (Frevel and Rogers 2016; Shearing and Johnston 2010; White 2012) are viewed as more agile, less costly and more responsive in resolving local safety and security issues than the old bureaucratic and centralized forms of governance. Thus, a strategy process in itself can improve interaction between stakeholders and enable to learn skills and create new knowledge. Strategic partnerships also both require and build mutual trust (Favoreu et al. 2016; see also Chap. 8). In a Canadian study people who expressed higher levels of confidence in the police were more likely to have accessed the police department's website or Twitter feed (Ruddell and Jones 2013).

Recent developments in information and communication technology (ICT) and mobile technology have made the Internet widely available. Simultaneously, social media have made their way into the everyday lives of citizens and also changed online communication considerably by bringing to it a new dimension: *interactivity*. Social media are popular, collaborative, participatory, decentralized and accessible in nature, making them a medium that allows information to pass quickly between multiple publics and organizations.

Together with smart mobile devices, social media have made creating, publishing and sharing of information, but also participation, easier and faster than ever. Empowered by these novel solutions, citizens have also come to expect public protection, safety and security authorities to use the same technologies.

Especially younger generations—equipped with more up-to-date smart devices—seem to have turned away from traditional media outlets such as radio, print and broadcast media towards social media such as Twitter as legitimate sources of information. Similarly, there are cases reported around the world where citizens have turned to modern channels such as Facebook, Twitter or even iPhone's FaceTime video session when seeking help or sharing information in an emergency situations. Public safety and security organizations have also increasingly used various social media tools in emergencies. (Akhgar et al. 2013; Reuter and Kaufhold 2018.)

Similar notions have been in the American Red Cross's (2011) study on social media in disasters and emergencies where more than 76% expect help to arrive within 3 h of posting their need to social platform. The trend of public sharing of information using wireless networks, mobile devices and social media, especially in mass events, can facilitate real-time two-way incident management (Atos 2014).

Coupled with these expectations for presence, prompt reaction, interaction and participation, the requirements for using multimodal content and varying services simultaneously challenge the traditional concept of official communication among public protection, safety and security authorities (cp. Chap. 2). Accounting for social media in communicating with the public without intermediaries expresses a big change in the official communication paradigm that consequently presents multiple internal challenges for public protection, safety and security authority organizations. Responding to these challenges, we have seen a plethora of research and developments projects, workshops, publications and seminars, with most of them focusing on the potential, possibilities and benefits of social media and mobile technologies but only few focusing on presenting *clear instruments for strategic or even organizational changes.*

Challenges of Social Media for Safety and Security Authorities

There are several aspects related to social media that affect the design of social media strategy. First of all, social media can be considered a complex ecosystem, with various dynamic and interconnected platforms: social networking sites (e.g. Facebook, Twitter or LinkedIn), user-generated content sites (e.g. YouTube, Wikipedia), game sites and marketing sites (e.g. Amazon, eBay) are interlinked and their boundaries blurred (Dijck 2013). This means that also the police presence is inevitably ***multidimensional*** and selecting few communication channels leads actually to simultaneous presence in many types of platforms with differentiated policies and cultures and various end-users.

Second, the use of social media among safety and security authorities presents also many *ethical and legal challenges*, most of which relate to the privacy of citizens involved. Information that the public authorities have and use is often sensitive. For example, acquiring or asking for private information from anyone involved in an emergency situation also raises questions of the contingencies in data management such as responsibilities for handling, managing, storing and using private information (photos of incapacitated people, addresses, phone numbers, location information, etc.) shared through any platform—either public or private (see also Chap. 3).

Thirdly, social media has also a *tendency to spread false information and rumours*. According to a study of Twitter feeds during crises, posts that do not mention the information source or include personal involvement and experiences are especially prone to become widely spread rumours (Oh et al. 2013). With respect to the findings of the EU Horizon 2020 projects iSAR+[1] and Soteria[2] (see iSAR+2019 and Soteria 2019), even though any additional information (photos, videos, location information, etc.) received is typically regarded as beneficial or useful for safety and security organizations, *information quality* has been regarded as troublesome. The information obtained from social media has to be processed before it can be utilized in emergency management activities. Based on the experiences and opinions of the authorities that took part in the iSAR+ and Soteria projects experimentations, use of social media increases the amount of useless or irrelevant information.

[1] **ISAR+ project:** Website date: 20 June 2016; project website: isar.i112.eu.

[2] **Soteria project:** Website date: 19 December 2016; project website: soteria.i112.eu.

Consequently, as the volumes of shared information increase, not only keeping track on what is going on in social media but also identifying false information and spam is becoming more and more difficult.

Hence, although social media can potentially enable monitoring citizens' and communities' fears and concerns, the enormous volume of social data streams generates "noise" that requires filtering and data mining in order to get meaningful information like major events or changes in public sentiment (Kavanaugh et al. 2012). As we have seen in many extraordinary cases such as the 2013 Boston marathon bombing, false or fake information can travel rapidly if they are spread through the right nodes (i.e. users with high numbers of followers; see also Chap. 7). Naturally this is dependent on the incident and the scale of it. Any information received through social media would need to be somehow validated before being utilized in situation management. This naturally has a great impact on resource allocation. The need for disseminating correct and up-to-date information increases in crisis situations. Efficient communication, informing and alerting are key in mitigating further damage.

However, in terms of peoples' trust towards different information sources, there are also conflicting notions in recent research. For example, Austin et al. (2012) have implied that traditional media outlets have been preferred as most credible source for information, with the belief that broadcasting companies and other journalistic institutions have the expertise and judgement in assessing and estimating an overview of matter to broadcast (Vihalemm et al. 2012). This is also supported by Reuter et al. (2016) study where only 13 % of respondents (1034 citizens from 30 European countries) considered information from social media to be more accurate than that presented in traditional media.

Accordingly, the role of social media has been seen more as an instrument to steer users towards appropriate information (Austin et al. 2012; Posetti 2012). Smart devices' and social media's features represent *media convergence* where different types of media and content—previously sectored behind different media outlets (radio, TV and print press)—are now merged together in one medium. Simplified access, creation, sharing and interaction to different content allows for more real-time information acquiring and awareness (See Chap. 10 for technical applications).

Implications of the Use of Social Media in Policing: Finnish Experiences

Overall there have been three purposes of social media in policing: (1) aiding in criminal investigations (e.g. to identify suspects), (2) disseminating information and managing public opinion, and (3) agency branding, reputation management and controlling virtual communities (Ruddell and Jones 2013). On the other hand, the interactive nature of social media places the police also under control. Control extends to both language and operational practices; camera and video recording features of mobile communication devices enable citizens to disseminate police practices and misconduct. According to recent research, consciousness of this has influenced significant behavioural changes, including moderation in the use of force among most officers (Brown 2016).

Virtual policing in Finland (specifically the Helsinki Police Department) started in 2008 when sergeant Forss and his team started preventive police work on Facebook, IRC-galleria[3] and Twitter (Virta 2013). Virtual policing evolved from a need to operationalize community policing in the Internet. Youngsters had started to communicate online using different kinds of applications and platforms. At the same time, there was a lack of police presence in social media. The task of visible online patrolling was to improve accessibility of police services and increase safety in the online forums where youngsters were spending their time. The presence created trust and confidence in the police and helped to receive information from citizens (Calcara et al. 2015; Virta 2013). The nature of using social media changed strongly when adults found it. Crimes and inappropriate conduct such as hate speech grew in number.

The next step of the Finnish virtual policing model was to develop three different methods of online policing: a nationally centralized Net Tip system (Nettivinkki), community policing in the Internet (Nettilähipoliisi) and a specific Internet Police (Nettipoliisi). *Nettivinkki* is an online channel for the public to give a hint of non-emergency issues, like drug use, hate speech or corruptive behaviour for the police. This

[3] IRC-galleria (IRC-gallery) was founded in 2000 as a photo gallery for the Finnish-speaking users of Internet Relay Chat (IRC). The site reports to have over 450,000 users, whose average age is 26 years (https://irc-galleria.net/). During the early years, the majority of the virtual community members were younger teenagers. Since then, other more global social networking sites like YouTube, Snapchat and Instagram have gained more popularity among the Finnish youth.

reporting can also be done anonymously. In addition, there is a possibility to officially report minor offences that do not require immediate police intervention online. Under the *Nettilähipoliisi* activity a group of local police officers use social media as a supportive tool for the mainstream police work such as preventive policing, crime investigation and maintaining public order and security (emergency response and patrolling). Preventive aspects contain networking and partnering with the local communities: giving advice and solving collaboratively their security problems. Local officers may also get in touch with people engaged in (online) demonstrations and build a dialogue before the assembly. Especially community policing in the sparsely populated areas with limited police resources benefits from social media tools. For example, one central Finland (Pohjanmaa) police force has developed their particular mobile community policing model to combine Instagram, Facebook and online chat as means of communication with the local public. The third way of working: the specific Internet Police (*Nettipoliisi*) is a continuation of the earlier virtual community policing method. Internet Police officers craft interactions between police and civil society through social media (Calcara et al. 2015). Nowadays every local police force has its own Facebook site and Twitter account on social media and dozens of personal specialist accounts on Twitter. The Finnish Police publishes also information of its awareness raising campaigns and other activities on Instagram and YouTube.[4]

The latest modification of policing on social media is a Virtual Operational Support team located in the police operational command centre of Helsinki. The team follows and analyses phenomenon on social media that are disruptive to societal cohesion. They provide reliable information in case of exceptionality and correct false information such as rumours, lies and hate speech provided by individuals or "troll factories". They support police patrols and investigation units in their daily duties and on the other hand provide webcare ("we know, we do, we care") to citizens in different languages. To enhance safety and security, the team enables a real time dialog with citizens on social media by being visible in most used platforms and apps including Facebook, Twitter and Instagram.

As described before, the strategic choice of being present on social media presupposes that all work processes and tasks are critically revised and—if necessary—updated. Sometimes the constraints of earlier working cultures may prevent achieving the strategic purposes and aims for using

[4] Links to different channels see: http://www.poliisi.fi/

social media as a new means of communication (e.g. for British experiences of the first years of engagement with the public on social media, see Crump 2011). When someone asks whether the police are effective on social media, it should be asked: effective at doing what? (Brodeur 1998). Due to the lack of evaluation studies, there are no proper measures to weigh the effectiveness of Internet policing. As all police work, also policing in the Internet is mostly performance management. So far, the evaluation has been done by counting the number of friends and followers in Apps and platforms or the volume of information exchange such as messages and emails (Calcara et al. 2015). More attention, however, should be paid to the distinction between assessing a process and outcome. The relevant question in the future is this: How are particular goals and criteria of success assessed in Internet policing?

A recent study of the possibilities for including social media as part of Finnish public protection, safety and security authorities' official communication concluded that the following aspects need to be addressed when adopting a social media strategy (Pylväs et al. 2015). Communication between the authorities and citizens in social media is mainly a one-way stream of information from the authorities to citizens, and largely based on pre-emptive informing and instructing. Actual *interactions* are still scarce, even though the value of social media in engaging the public has been recognized. This was seen partially due to the organization of communication activities as specialized units within public protection, safety and security authority organizations. Social media use is typically tied to appropriate positions in terms of skills, responsibilities and in some cases personal interests, which in turn has led to an uneven distribution of responsibilities with respect to social media communication and uneven resources to attend to related communication activities.

Yet attitudes towards social media use were generally positive; the respondents saw clear potential in using social media as a communication tool. Actual guidelines, instructions and training were, however, considered to be missing. Legislation was not seen as an obstacle per se, but rather as restricting; insufficient awareness of privacy and data protection law's applicability in terms of social media communication was considered resulting in social media communicators being overly precautious and refraining from interacting with the public. As a result, the respondents within the study proposed several suggestions to address these challenges:

- Official communication should be as uniform as possible.
- Communication—in general—should be viewed as a more integral part of everyday police work.
- Social media should be recognized as one of the official communication forums.
- Education and training in public communication should be increased, both within basic training and further training (continuous education).
- Communication activities should engage the public in interaction.

Service Design Concept for Policing in the Social Media

During the EU Horizon 2020 project Unity,[5] the team at the Police University College (Finland) developed on the basis of a business model canvas a strategy tool named a *service design canvas for community policing* (SDCCP; see Fig. 4.1).[6] The tool is for addressing and prioritizing the needs of communities and citizens, setting goals while recognizing how various resources, activities, partners, resources and channels may have an effect on the outcomes. *Service design* can be used as an overall strategy tool for community policing, or as a more specified tool for example in designing a social media strategy.

There are nine elements in the SDCCP template. Contents for these can be identified with the assistance of SWOT, PESTLE or other environment analysis methods. The template can be first filled by police and later be turned over to already identified strategic partners or even beyond for critical assessment and feedback for improvements. The first five elements in the template are about the operational environment of policing. Obviously, these elements receive material and ideas from the critical examination of opportunities and threats in SWOT or similar environment analyses.

[5] The fundamental vision of the Unity-project (2015), funded by the EU Horizon2020 program, was to strengthen the connection between the police and stakeholders to improve the safety and security of communities and citizens. Besides enhancing community policing practices, UNITY has developed technological tools for communication and collaborative problem solving and piloted these tools in eight European countries. For more see CORDIS, EU Research results platform: https://cordis.europa.eu/project/rcn/194893/factsheet/en

[6] The original business model canvas is distributed under a Creative Commons license from Strategyzer AG and can be used without any restrictions for modelling businesses.

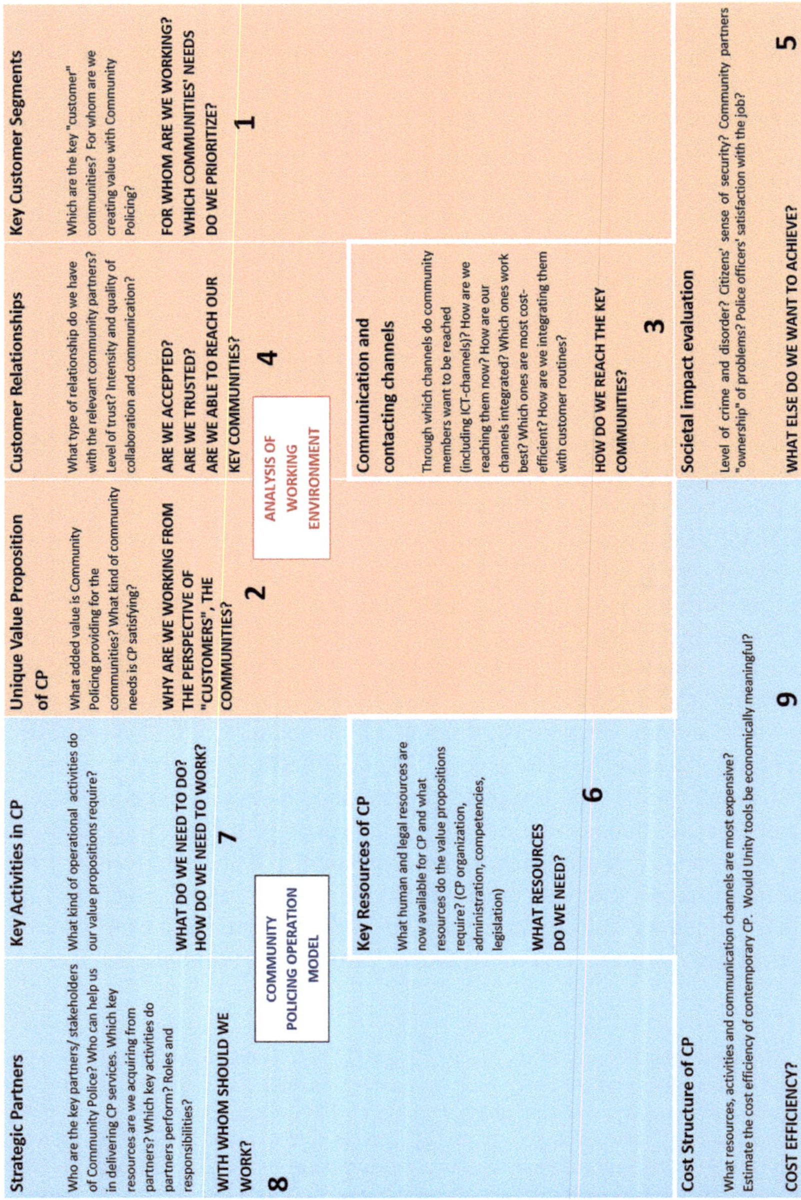

Fig. 4.1 Service design canvas for community policing

At the beginning, it should be defined, who are the *key customers* of policing. Which (virtual) communities, minority groups, individuals or localities will be beneficiaries of the services? For whom is police creating value with their presence in social media? This requires thorough discussion and judgement, because not all customers can be reached via social media, at least with the tools available for police. The next phase requires imagination and innovation. The participants of the strategy process should put forward a *unique value proposition* for policing. What kind of added value is policing in social media providing for the key customers? What kind of security or safety needs and expectations should policing meet?

The third element in the SDCCP is the identification of the main *communication and contacting channels* with the key customers. Through which social media channels do key customers want to be reached? Channels could be various types of ICT tools such as platforms usable by PCs or smartphones but also the more traditional arenas such as community meetings and the use of trusted intermediaries could be added here as supportive channels. Social media should not be seen as a separate form of policing. As already mentioned, for example, in Finland a particular model of mobile community policing for sparsely populated areas has been developed, where the patrolling officers meet citizens face-to-face and can be reached via private web chat and social media platforms (Instagram, Twitter and Facebook).

It is useful to consider pros and cons of the ways and tools that are used currently for approaching various community members and individuals. Additionally, one should review how the various communication and contacting channels are integrated together; for example will there be a joint platform for email, social media and the Internet? Which channels have worked well so far and are likely to work well in the future? It is also important to consider the aspects of security and cost-efficiency of various communications options, if digital services will be considered for communication and repository.

Fourthly, the users of the service design canvas should analyze the quality and intensity of all significant *customer relationships*. Several questions are relevant here. How would you describe the relationships with the different customer groups and strategic partners? How would you like to change these relations in the future? How strong is the level of trust between different parties? What is the current intensity and quality of collaboration and communication between the parties? How well are the police and its practices accepted by different (virtual) community members? Are the police able to reach all the key (virtual) communities, especially minorities and groups or individuals in marginal or weak positions? Why are some difficult to reach? The final element of the operational environment is the evaluation of the wider *societal impacts* of the social media presence. What are the benefits? How should it impact reported crime levels or citizens' perception of crime and feeling of safety? How would the police officers feel about their work and working conditions?

After recognizing the needs for security and well-being needs of key customer communities, one is ready to start designing an operational social media policing model. This comprises defining the key activities, key resources and strategic partners and finally making a cost structure assessment. Then the sixth phase in the service design canvas is to define the *key activities* in community policing. What kind of operational activities do the value propositions require? In other words, what do you need to do, and how do you need to work?

The next step is defining the *Key Resources* available for policing in social media. What human, technological, organizational, economic, administrative and legal resources are now available and what resources do the value propositions require? Resources can also include various kinds of competencies, which broadly defined consist of skills, knowledge, motivation and values in addition to opportunities to put those skills and knowledge to work for the service of (virtual) communities.

The eighth phase looks back at the key customers and considering which stakeholders or individuals would be the most critical partners in creating value. Who are the *key strategic partners*? Who can help in delivering policing services in social media? Which key resources can these partners bring with them for the benefit of the whole virtual network? Which key activities could the partners perform? What would be their specific roles and responsibilities in the collaboration?

The final element in the service design canvas is the analysis of the *cost structure* of policing with the defined resources, partnerships and activities. What are the costs for alternative resources, activities and communication channels? Estimate the comparative cost efficiency of current policing model and the planned model including social media presence. Costs and benefits can also be calculated for alternative solutions to particular security problems.

The ultimate goal in using the service design canvas for community policing is developing value for and with the customers, or rather partners of police. The strong point of SDCCP is that it can be outlined on one page. The outline can be done together with the relevant stakeholders, and if necessary be tested and assessed with another group, if they cannot be engaged with the design process for some reason. The feedback from stakeholders and the (virtual) community members can help assess whether the chosen service design strategy is desirable. The first few social media interventions may not be perfect, but it is important to tolerate mistakes and allow collaborative learning.

Conclusions

Bueerman (2012) predicts that in the future police will be more and more a broker and a facilitator for community actions to solve problems related to crime and disorder. The role of citizens and civil society will change from mere information receivers and security consumers to partners and co-producers, who will take their local knowledge, skills and resources to the production of security services. Social media tools can be a part of this operational change if the police organization and communication practices are adjusted accordingly to support more interactive and collaborative policing.

As security becomes a public good that is co-produced through various networks and actors, the police need a better understanding of social media and what kind of added value it can bring to policing. This chapter aimed at offering the police and their partners such an understanding by suggesting one tool for improving policing services. The service design canvas could assist in constructing the social media strategy, to build safety and security not only for but with the communities and citizens.

References

Akhgar, B., Fortune, D., Hayes, R. E., Guerra, B., & Manso, M. (2013). Social media in crisis events: Open networks and collaboration supporting disaster response and recovery. In *2013 IEEE International conference on technologies for Homeland Security (HST)*.

American Red Cross. (2011). Survey Data: Social media in emergencies. Retrieved from: https://www.scribd.com/presentation/63022972/SURVEY-DATA-Social-Media-in-Emergencies-2011

Atos. (2014). *Safeguarding your citizens and assets with emergency management.* Retrieved from http://atos.net/content/dam/global/documents/your-business/atos-emergency-management-whitepaper.pdf

Austin, L., Fisher Liu, B., & Jin, Y. (2012). How audiences seek out crisis information: Exploring the social-mediated crisis communication model. *Journal of Applied Communication Research, 40*(2), 188–207.

Brodeur, J.-P. (1998). *How to recognize good policing, problems and issues.* London: Sage.

Brown, G. R. (2016). The blue line on thin ice: Police use of force modifications in the era of cameraphones and Youtube. *British Journal of Criminology, 56*(2), 293–312.

Bueermann, J. (2012). Preparing the police for an uncertain future: Four guiding principles. In D. R. Cohen McCullough & D. L. Spence (Eds.), *American policing in 2022: Essays on the future of a profession.* COPS, U.S. Department of Justice. http://ric-zai-inc.com/Publications/cops-p235-pub.pdf

Calcara, G., Forss, M., Tolvanen, M. J., & Sund, P. (2015). The Finnish internet police (Nettipoliisi): Towards the development of a real cyber police. *European Journal of Law and Technology, 6*(2), 1–14.

Crump, J. (2011). What are the police doing on twitter? Social media, the police and the public. *Policy & Internet, 3*(4), 1–27.

Dijck, J. (2013). *The culture of connectivity: A critical history of social media.* New York: Oxford University Press.

Favoreu, C., Carassus, D., & Maurel, C. (2016). Strategic management in the public sector: A rational, political or collaborative approach? *International Review of Administrative Sciences, 82*(3), 435–453.

Frevel, B., & Rogers, C. (2016). Community partnerships (UK) vs crime prevention councils (GER). *The Police Journal: Theory, Practice and Principles, 89*(2), 133–150.

Hoogenboom, B. (2010). *The governance of policing and security – ironies, myths and paradoxes.* Basingstoke: Palgrave Macmillan.

Kavanaugh, A. L., Fox, E., Sheetz, S. D., Yang, S., Li, L. T., Shoemaker, D. J., et al. (2012). Social media use by government: From the routine to the critical. *Government Information Quarterly, 29*(4), 480–491.

Oh, O., Agrawal, M., & Rao, H. R. (2013). Community intelligence and social media services: A rumor theoretic analysis of tweets during social crises. *MIS Quarterly, 37*(2), 407–426.

Pawson, R., & Tilley, N. (1997). *Realistic evaluation.* London: Sage.

Posetti, J. (2012). The Twitterization of ABC's emergency and disaster communications. *The Australian Journal of Emergency Management, 27*(1), 34–39.

Pylväs, K., Hokkanen, L., & Kankaanranta, T. (2015). Sosiaalinen media ja mobiiliteknologia avuksi viranomaisviestintään. *Sisäministeriön julkaisu* 7/2015. Sisäministeriö.

Reuter, C., Ludwig, T., Kaufhold, M.-A., & Spielhofer, T. (2016). Emergency services' attitudes towards social media: A quantitative and qualitative survey across Europe. *International Journal of Human-Computer Studies, 95*, 96–111.

Reuter, C., & Kaufhold, M.-A. (2018). Fifteen years of social media in emergencies: A retrospective review and future directions for crisis informatics. *Journal of Contingencies and Crisis Management, 26*(1), 41–57.

Ruddell, R., & Jones, N. (2013). Social media and policing: Matching the message to the audience. *Safer Communities, 12*(2), 64–70.

Shearing, C., & Johnston, L. (2010). Nodal wars and network fallacies. *Theoretical Criminology, 14*(4), 495–514.

Unity-project. (2015). *Unity—Creating a new, community-centered approach to cimmunity policing*. Retrieved from https://www.unity-project.eu/wp-content/uploads/2015/12/unity_Brochure.pdf

Vihalemm, T., Kiisel, M., & Harro-Loit, H. (2012). Citizens' response patterns to warning messages. *Journal of Contingencies and Crisis Management, 20*(1), 13–25.

Virta, S. (2013). Finland. In M. K. Nalla & G. R. Newman (Eds.), *Community policing in indigenous communities* (pp. 247–256). Philadelphia: Taylor & Francis.

White, A. (2012). The new political economy of private security. *Theoretical Criminology, 16*(1), 85–101.

CHAPTER 5

Public Crime Reporting on Social Media: A Progressive or Regressive Phenomenon?

Alice Raven

Introduction

Visual criminology has become an expanding sphere within cultural criminology, which can be applied to exploring the emerging phenomenon of the public using social media to report crime incidents (Sandberg and Ugelvik 2017). With links to community surveillance and news reporting, public crime reporting can be understood as a cultural phenomenon that emerged on the background of technological and cultural developments, which reshaped the conventional relationships between crime and imagery. Within a globalised technological community, individuals have nuanced capabilities of recording and disseminating crime information on social media from a first person perspective, taking the identity of 'community news reporters' that can provide online audiences with real-time evidence into a crime situation (Bock 2016). Antony and Thomas (2010) argue that this has succeeded the traditional inaccessible and highbrow news networks that dominated early social media frameworks where a crime was recorded and reported predominantly by the professional news agencies and disseminated to a widespread audience within the commu-

A. Raven (✉)
CENTRIC, Sheffield Hallam University, Sheffield, UK
e-mail: A.Raven@shu.ac.uk

B. Akhgar et al. (eds.), *Social Media Strategy in Policing*,
Security Informatics and Law Enforcement,
https://doi.org/10.1007/978-3-030-22002-0_5

nity (Bock 2016). The new method of documenting crime on social media has facilitated a cultural shift from traditional reporting standards, which reflects a change into how the public responds to emergency situations—which has transitioned from contacting the emergency services to capturing the scene using a smartphone and disseminating the content on social media.

The technological advancements of modern society and the acceleration of social media platforms as a part of everyday activities have collectively proven themselves as a facilitator of change for how crime is reported. To understand the cultural shift in its full context, it is important to highlight the traditional framework of crime reporting online to reveal how the public has transitioned from passive receivers of information to what Stokes (2013) and Bock (2016) label 'media fans' and 'activists' who actively engage in crime reporting online. They add that the first wave of crime reporting can be understood as a single, direct channel that flows from the smaller proportion of 'media producers', who create and disseminate crime reports, to a large online audience of 'listeners' that absorb the information. This early interpretation of crime reporting as a one-way communication system and arguably depicts the public as an accepting, passive audience that are predominantly consumers of information that is widely agreed as accurate and reliable. The transformation of the traditional crime reporting structure into a two-way flow of information was heavily influenced by the emergence of smartphones and social media advancements (Yar 2012). Yar (2012) argues that the emergence of smartphones into the modern market has become the forefront of mediating public interactions, as individuals have a direct connectivity between the camera lens that is present at the crime incident and the homepage of social media sites, which allows them to create and disseminate real-time information. Stanley (2014, 2017) labels this technological acceleration as 'Little Brother' surveillance, which allows the public to broadcast mundane interactions amongst themselves and other to a wider audience. In relation to crime reporting, these developments and innovations have given the public a sense of freedom and means of using the camera lens as a 'third eye' to personally document crime-related incidents (Bock 2016; Mills 2003).

This cultural change has led to what can be understood as the 'second wave' of online crime reporting which has become fully immersed within modern societies, where large institutions and news agencies are no longer the predominant producers of crime information. Simultaneously, the public are no longer confined to being apathetic consumers but have become a subculture of producers that generate crime reports through imagery (Yar 2012). This growth in individual crime reporters that are using social media for its constant interconnectivity are defined by Yar (2012) as the nuanced hybrid identity of 'prosumers', acting as both producer and consumer of crime related content. Stokes (2013) comments that this term reflects the replacement of text with imagery as the main source of news communication, for its ability to present audiences with a 'first person' view of a crime incident.

Reflecting on this cultural shift in crime reporting, understandably this phenomenon is on the rise with social media replacing law enforcement as the quickest method of disseminating and reporting crime information. Miller (2011: 171–173) supports this arguing that social media enables individuals to document viewpoints and experiences through the eye of the camera—an emerging part of popular culture labelled as 'microblogging'. This idea of continuous documentation of individuals' personal lives reflects the dissolving boundaries between the public and private sphere, with the originally 'hidden' experiences of individuals increasingly visible on a global scale to create a prominent social presence. In this new crime reporting dynamic, the public prosumers have harnessed social media to replace the traditional understandings of how 'reporters' of crime incidents are perceived. From witnessing crime incidents, the public are using the camera as the third eye, the smartphone as the direct link to their online social media profile and the social media sites to disseminate their content to a transnational audience of fellow prosumers (Mills 2003).

Community surveillance has provided an enriched and unprecedented new sphere of exploration into understanding crime reporting, and how social media and technological developments have created a globalised space for news dissemination. This chapter will delve into the cultural criminological concept of visual criminology and apply its theoretical frameworks to explore public crime reporting online through three parts. First, the chapter will discuss the perspectives and identities that are recognisable from reports, which reflects upon themes of victimology and the motivations of crime reporting. In this context, it will discuss the identity disparities between crimes recorded by the 'reporter' recording others as a witness to external 'offenders', and more abstractly discuss the rise in a new perspective of reporters recording their own crimes as offenders and how these perspectives are shaping the future of crime reporting.

The second part will take inspiration from the motivations of crime reporting to explore an emerging academic debate that discusses, through two contemporary case studies, whether crime reporting on social media is a 'just' and acceptable act. This debate will be discussed with regard to the phenomenon's cultural and social impact, that public crime reporting reflects a form of social activism which represents cultural and social progression and the opposing critique that crime reporting is a misunderstood perspective that—through unjust practices—reveals the hidden framework of an unattached and regressive society. The exploration into how crime reporting is produced through perspectives, and the growing debate surrounding the 'justness' of its practices, reveals the importance of researching the phenomenon's intertwining themes, which are woven into its cultural infrastructure.

Providing a more constructive approach the penultimate section will reflect on how public crime reporting on social media has been used by law enforcement agencies to enhance investigations, using several examples to show current best practices. To conclude, the chapter aims to provide an in-depth discussion into the perspective features woven into the cultural framework of public crime reporting, the emerging debate surrounding its social and cultural implications and to display how this new phenomenon can positively transform the current methods used by law enforcement to enrich intelligence for investigations.

The Perspectives of Reporters

The impact of social media on crime reporting by reporters can be understood through three perspectives, which help to unravel the motivations behind public crime reporting on social media. The perspectives focus on the reporter, or public member, and their self-proclaimed association with the crime situation. The most common perspective, which is widely recognisable on social media, is the reporter identifying themselves as a 'live' witness of a crime providing online audiences with a real-time, first person capture of the crime. Sandberg and Ugelvik (2017) argue that in contemporary society the phenomenon has increased the urge for 'witnesses' to use their smart phones to record a crime incident to the extent that it has become a mundane response. This form of public crime reporting has created a shift towards a new form of 'narrator' that is the first person at the scene and in some instances has an in-depth knowledge of the context surrounding the crime. This has ultimately removed the need for 'eyewitness' statements that were traditionally used to describe crime incidents, as witnesses are now using the camera lens as their 'eye' to allow audiences a clear view into their viewpoint of a crime incident. Reporting of this kind still has remnants of traditional crime reporting methods, such as the incorporation of verbal commentary describing the incident alongside a written overview, which provides social media viewers a detailed eyewitness statement of a crime incident.

A feature of traditional crime reporting that can be seen in the contemporary phenomenon is the type of 'victim' that is labelled by the reporter in the report. This can be understood using Chris Greer's (2006) analysis of Nil Christies notion of the 'ideal victim', a socially constructed identity that defines the parameters of what constitutes a 'true' crime victim. This hypothetical identity is interwoven with predetermined hierarchal values, which prescribe the characteristics of a victim's 'deservingness' based on visualisation and context. Consequently, when public members witness a crime, they use these predetermined characteristics to make a judgement of whether the individual deserves the victim status. The comparison of the victim is made against these characteristics (e.g. white, female, very old or young), in order to determine their level of victim status (Greer 2006).

Once this identification of victim and offender has been decided by the reporter and shared on social media, the labels become 'fixed' and reinforced as 'factual' by the online community. Often this can have a profound effect for the victim, which draws attention to Becker's (2008) research that a 'victim status' becomes a 'master status' and consequently a stigma that in many cases can become difficult to dissolve. Noticeably, what does challenge traditional understandings of the ideal victim is the increase in witness reports circulating on social media that incorporate victims from ethnic minority backgrounds that defy the archaic victim characteristics. Activist groups with large social media followings such as *Black Lives Matter* (2019) regularly share and repost public crime reports that include an ethnic minority victim, which reinforces the idea that the parameters of the 'ideal' victim are expanding as activist groups are redefining values that underlie traditional victim frameworks.

Studying the witness perspective of public crime reporting reveals a larger network of cultural and societal hierarchies that indirectly influences the digital identification of victims. The notion also shows how contemporary social activism represents a motivation for public crime reporting by challenging and reimagining the parameters of the 'ideal victim', replaced by a more dynamic and inclusive framework influencing online cultural change.

Another well-established perspective that aids current understandings of the motivations behind public crime reporting is based on the identity of the reporter as a victim. The phenomenon has manifested a new form of 'victim perspective', which gives the online audience a direct insight into a victim's account of a crime incident. What is interesting from this perspective is that the reporter immediately associates themselves as 'deserving' of the victim status and posts their report on social media with this self-identification label intact (Greer 2006). Consequently, when the online audience views the report, they will be supplied with a preconceived narration of who the victim is in the situation.

In exploring this concept further there are two distinct methods used by victim reporters to record the incidents, either in form of a 'selfie' or 'first person' stance. Eler (2017) argues that recording a crime from a *selfie stance* shows similarities to that of news reporters, where the producers use their front-facing smartphone camera to record their face which can include commentary and a direct conversation with the online audience. He adds that this perspective reflects an interconnectedness between the 'public' and 'private' spheres, as public members are merging the two through imagery to create an online identity that is shared within the online community. An increase in this 'selfie' crime reporting is identifiable with domestic abuse victims, who create a 'timeline' of abuse images through the form of selfies labelled as 'Survivor Selfies' and who document them on social media as an awareness raising technique (Domestic Shelters 2014; BBC News 2017). Notably this shows how social media is perceived as more than simply a place to report crime; rather it is seen as an interactive community which encourages social activism.

On the opposing side of the smartphone, the backwards facing lens facilitates the *first person victim perspective* whereby the camera symbolises the reporters 'third eye' (Bock 2016; Mills 2003). This vantage point gives the online audience a 'mirrored' view to the reporter, which displays their outlook of events. A sphere of society, which has adopted this use of crime reporting particularly frequently, is victims of hate crime, who use their camera phones to record and disseminate the content online to again spread awareness and protest. Lombardo (2019) argues that this method has been used by members of the LGBT+ community, who remain a large target for this type of crime particularly in the UK. He adds a recent example including a young LGBT+ couple that were assaulted on a bus in London and that posted an image of themselves with blood stained clothes on Facebook in response. The post received over 10,000 comments and was widely acknowledged on similar platforms such as Snapchat and Instagram (Lombardo 2019). The sheer breadth of online audiences can explain why individuals are motivated to report crimes on social media, particularly for the purpose of awareness raising. The examples show how smart phone features are used by individuals to create the victim focused perspective and reveal its ability to facilitate and impact social and political discussions throughout the online community.

What becomes interesting in researching the perspectives of social media reporting online is how the boundaries between reality and the producer's perception of reality, and what is factual, become blurred (Sandberg and Ugelvik 2017). The reliability of the public member as a witness or victim provides an interesting sphere of research, as imagery is often falsely labelled by the wider community as 'factual evidence' that only captures the truth of a situation. Pande (2017) argues this misconception stems from the absence of a standardised ethical guideline to regulate public journalists on social media, which for journalism agencies was developed and disseminated by the professional journalism sphere to promote best practices. Antony and Thomas (2010) add that the absence of traditional regulations removes the requirement for public producers to adopt journalistic values such as withholding a value judgement when creating and disseminating content online. They add therefore that social media has become a dystopian reality, where platforms are continuously filled with unregulated content that distorts the public's perceptions of the context surrounding crime, including who the 'person to blame' or the offender is in the situation (Office for National Statistics 2019). The misperceptions of crime incidents can lead to gender and regional biases and stigmas held by the audience, which discriminates individuals as offenders and reinforces cultural conflicts which further oppresses and marginalises subcultures (Pande 2017). As a consequence, many experts have begun to adopt an 'anti-public journalism' ideology due to the lack of ethical guidelines has led to much criticism concerning the unreliability and trustworthiness of publicly generated content (Antony and Thomas 2010).

Perhaps a more abstract and upcoming perspective, which has significantly impacted how crime is reported online, is the manipulation of technological advancements by offenders who produce and disseminate their own crime reports. Sandberg and Ugelvik's (2017) cultural criminological framework helps to explain why offenders have begun using smartphones to record and disseminate imagery of themselves committing offences.

This new form of crime reporting has facilitated an interconnected web of extremist imagery that forms a visual culture, often removed from the mainstream online audience. Members of extremist groups are often heavily embedded within this visual culture and record their crimes as a way of contributing towards the wider snapshot culture (Solon 2017). The viewpoint extremist producers draw upon in particular is influenced by gaming culture, using body cameras as the role of director to recreate a 'first person shooter' type simulation. The Christchurch shooting in New Zealand provides a prime example, in which the offender used the Facebook live feature to create a 'real-time production' incorporating video and music which garnered over 4000 views before it was removed (BBC News 2019). The 'live' feature provides offenders with a direct link to their social media audience, on which they can share the real-time experience of the crime unfolding to an engaged online extremist community (Solon 2017). This nuanced trend of crime reporting reflects a darker turn towards using social media as a platform to disseminate extremist content that is becoming commonplace especially during terrorist attacks.

Using sociological and criminological frameworks to explore the number of dynamic perspectives and identities that are entrenched within the public crime reporting process reveals an underlying set of motivations. Through further inspection of these viewpoints a number of conclusions can be made. The development of the public crime reporting has provided a new framework of perspectives for understanding visual criminology. These frameworks directly influence and challenge the conventional understandings in defining and remoulding how victimology in news reporting is understood. Furthermore, it is these perspectives which reflect the ongoing debate that critically examines the reliability of public crime reporting and how disseminating publicly generated data online can create false narratives and accelerate cultural conflicts. What can be observed is a growing debate circulating across the criminological sphere, which discusses whether the motivations behind public crime reporting influence its portrayal as a 'justified' act in modern societies. These conclusions reflect a nuanced research sphere focused on understanding criminal behaviours and motivations in the social media age.

The Debate Surrounding 'Just' Crime Reporting

To fully engage with and understand the rise in public crime reporting, it is important to explore beneath the surface of common academic and wider public perceptions of the phenomenon. Importantly, scholars have identified the need to uncover the underlying semantics within cultural, political and social values that directly influence individual's motivations to report crime alongside the societal implications that may arise as a result. Recurring within the literature is the debate, whether public crime reporting on social media is a 'just' act. From one perspective, the phenomenon reflects a positive use of social media as a form of activism that outlines modern social injustices, whereas conflicting literature argues that the common practice of social media reporting has led to the emergence of the 'bystander' phenomenon. This debate will be outlined using two contemporary case studies that each present an opposing ideology, which through examination will either challenge or complement current understandings.

The first study explores how a network of student subcultures have emerged online following the positive use of social media as a social activist platform to circulate their perceived institutional betrayal of universities concerning sexual abuse cases. The second case study discusses how public crime reporting on social media has consequently influenced the emergence of the 'bystander effect', which presents a rise in unethical recording of crime victims. The comparative analysis of the two cases will present an in-depth examination into the contemporary debate surrounding public crime reporting as a just practice.

Crime Reporting as Social Media Activism: Gender-Based Violence at Universities

Many scholars have discussed the hegemonically dominant cultural and institutional values, which indirectly influence gender-based violence in University communities but have not yet studied its connection to the upsurge in social media crime reporting. The transition of University institutions from a place of personal empowerment to a place of concern regarding sexual abuse and harassment has gained increased attention from the media, academia, activists and the wider community (Lewis et al. 2018). This stems from the publication of contemporary research such as the British national survey on *Violence Against Women* Studies (2016) and the publicly accessible legal action cases in the US that involved 100 colleges (Lewis et al. 2018). Busby (2019) adds that cases similar to these, particularly in the UK, revealed the sheer scale of abuse that based on an underlying 'lad culture'. Sexual health charities such as Brook revealed that more than half of the surveyed students had experienced unwanted sexual behaviour, but only 8% of the cases have been reported to institutions (Busby 2019).

The research published begs the question, why not more victims report their experiences to institutions. Lewis et al. (2018) suggest this is due to the perception of 'institutional betrayal', a nuanced phenomenon created as a result of victims experiencing difficulty in reporting sexual abuse, consequently leading to a breach in trust with university institutions. Universities have been increasingly criticised by students for their insensitive and undermining approach to sexual abuse reports, by having limited psychological well-being support, ineffective reporting procedures and not working cooperatively with other universities to tackle the issue (Somerville 2019). Oppenheim (2019) argues that these factors have created a 'university bubble' in which victims and offenders will socialise, work, study and live within the same parameters. It is these growing tensions and frustrations of students towards universities, fuelling institutional betrayal perceptions, that has led to an increase in students turning to social media to report their experiences of crimes as sexual abuse victims.

Social media has become a platform for university victims of sexual abuse to document, report and share their experiences of a crime. Oppenheim (2019) argues that social media acts as a 'safety mechanism' for victims, as it provides a network of support outside of the 'university bubble', allowing victims to raise awareness of sexual abuse at university. She found that students at Leicester University, for example, used social media to share their experiences of sexual assault on campus, an act which they believed represented a sense of empowerment and social presence to inform and raise awareness to an online audience. What this article emphasises are the disparities between the support given to sexual victims online and offline, which explains why many students and young people are in favour of reporting crime experiences to a more actively engaged audience on social media.

Oppenheim (2019) found that the most recognisable audience was the wider student community, who rallied and encouraged victims to issue formal complaints. This study hints at the creation of a new online student subculture through crime reporting, which was created from a shared distrust in universities and knowledge of social injustice. Lewis et al. (2018) argue that this subculture was formed as a branch of a new young feminist subculture, in which universities represent the 'hotspots' of its growth. By joining the subculture, students are given a chance to resist the underlying oppressive university culture through feminism and collectively strive to actively change their futures.

An example that demonstrates this cultural progression online is *It Happens Here* (2019), a student-based movement created by Oxford University students to raise awareness of sexual violence on campus. They have a strong social media presence that spreads across major platforms such as Facebook, Twitter and YouTube. The platform allows victims of sexual abuse to anonymously report their experiences including how their universities responded. This subculture reinforces the transition of crime reporting online by building upon the institutional betrayal of universities, a value which has coincidently globalised across the UK and into the US where surrounding university subcultures have emerged in solidarity to the movement.

The global impact of these subcultures—built upon the growing tensions between students and universities—can also be seen in *#StandUpJapan*, a social movement created following the release of an article which ranked universities based on how sexually promiscuous the female students were (Yamasaki 2019). These cases show how crime reporting has become embedded with gender-based social and political tensions, where the creation of globalised subcultures have accelerated the transition of young people reporting sexual abuse on social media. Lewis et al. (2018) argue that these subcultures have significantly influenced the re-evaluation of archaic university norms, developed new policies and the strategic development of universities, an impact they otherwise would struggle to achieve offline in the current climate.

Together, these cases highlight the correlation between the growing (awareness of) institutional betrayal and the rise in young people using technology and social media as a just method of reporting sexual abuse crimes. This transition has consequently led to the emergence of a number of student subcultures that are reinforcing this change and providing an accepting space for victims to report crimes; a space that has yet to be introduced into university communities. Badalge (2017) supports these conclusions arguing that witnesses are adopting the role as social activists, who although they are not physically intervening in crime incidents, they are creating social upheaval within the online community. This case study reflects on the debate, showing that online crime reporting can support social activists as a just method.

The Rise in Public Crime Reporting and the 'Bystander Effect'

Analysing how the public responds to emergency crime situations, it appears that individuals' reactions to an observed crime has changed significantly. By exploring the research into the phenomenon, the growing concern of social media's negative cultural and social influence become more recognisable. The common practice for the public to respond to crime incidents by raising their smart phone camera to directly report the incident to social media has been labelled as a regressive aspect of society (Uelmen 2017). This socially constructed phenomenon of online crime reporting was coined by Darley and Latané (1968) as the 'bystander effect'. Investigating why witnesses of crime incidents are behaving passively, (instead of intervening or calling for help) they identified two factors. When individuals approach a crime incident where a crowd is present, they experience a decreased level of self-responsibility based on the belief that another spectator will intervene (Badalge 2017). Tatham (2016) adds that human beings are also rational actors, i.e., on approaching the crowd individuals will assess the behaviour of others who will socially influence the decision-making process. If the individual believes that the members of the crowd are more knowledgeable of the situation than themselves, they will follow the actions of the crowd by remaining a bystander from the preconceived belief that the situation is under control (Darley and Latané 1968).

An interesting aspect in transferring the research on the bystander effect and its influence on public crime reporting on social media is the tension between the individual and the wider community regarding their just motivations for reporting the incident to social media. Members of the wider community may believe that public reporting reflect a growing number of social spectators who treat crime incidents in a similar manner to theme parks (Uelmen 2017). Badalge (2017) applies this concept to the Malaysian Airlines incident in 2017, where he describes the witnesses as a 'silent audience' who—instead of intervening—fed a live stream to the social media audience with multiple videos and images of the scene, causing global public outrage. The transition from active interveners to passive bystanders using social media to report crimes has led to the development of what Uelmen (2017) labels an online community that reinforces the bystander effect by encouraging users to record crime instances and share them online. He turns to the social media group *WordStarHipHop* as an example to showcase the popularity of these communities; *WordStarHipHop* has over three million visitors per day. This differs largely from the 'social activist' perspective on the debate, showing how public crime reporting online is being conducted for more regressive reasons by creating spectator environments that distorts crime reporting's progressive nature.

Hill (2016) argues that these socially constructed communities are able to sustain themselves by encouraging and relying upon the public to update the sites with self-generated crime imagery, thus replacing law enforcement as the place to report crime. Research into these new online communities explains why the bystander effect has become commonplace and how social media maintains its role as the new sphere of crime reporting. This section outlines the two sides of the debate: the one that shows that the public are responding 'unjustly' to the situation by not intervening and the other showing that perhaps the new phenomenon of social media crime reporting represents a new form of intervention.

To explore the critical perspective of the debate whether crime reporting online is just, examining the public's response to recent terrorist attacks proves an interesting case. There seems to be an increase of bystander-focused crime reports created, uploaded and circulated around social media by the public following terrorist attacks. This has occurred during several attacks such as in Westminster, London (2017), where witnesses recorded and took images of the victims which were then uploaded online (Botsford 2017). Through analysing social media posts following the Westminster attack, Botsford (2017) highlights the public's concerns of the bystander effect; e.g. "really disagree with some of the photos … published of the Westminster victims, beyond disrespectful." Botsford (2017) adds that this is not an entirely new phenomenon as there were also instances of public bystanders following the devastating crash in August 1997 in Paris involving Princess Diana of Wales. Cases similar to the ones in London and Paris reveal that the bystander effect has become a globalised behaviour, which raised critical discussions of how the public reporting of terrorist incidents reflects an unjust practice.

Drawing on victimology studies, Uelmen (2017) highlights the cultural implications of the phenomenon arguing that in recording terrorist incidents the victims become objectified, as public members dissociate themselves from the scene and exploit their privacy rights without receiving any form of consent. Badalge (2017) labels this objectification as 'unethical witnessing' which has led to a rise in digital frustrations. Hill (2016) comments that this regressive practice is a result of a dissolution of empathy with the victims, who are stripped of their 'human identity'. Alternatively, Hill (2016) adds that the onlookers (labelled 'gawkers') do not consider who is watching within the online audience. In the context of terror attacks this raises the issue that—due to the direct stream between the bystander's camera and social media—family members and friends of the victims may unexpectedly discover the information online as opposed to being informed by law enforcement. This raises the question from a law enforcement perspective of whether this new phenomenon has become obstructive and impedes their efforts following crime incidents.

Reviewing the criticisms surrounding public crime reporting online has highlighted the phenomenon as an unjust and regressive practice that encourages witnesses to become un-empathetic bystanders resulting in growing cultural division.

Exploring the contemporary debate into whether public crime reporting represents a positive or negative influence on the social, cultural and political spheres in an interconnecting online and offline community illustrates a clear disparity within society of what is the most 'just' response to reporting crime incidents. There is a clearly conflicting set of ideologies, in which social media was identified by some as a positive space for the public to take the role of news reporter and inform the online audience about social injustices as a form of social activism, while critics have argued that this had led to the development of an unattached and passive society that has reinvented how crime incidents are responded to by the wider public.

The different cases have shown how the phenomenon encompasses both viewpoints, highlighting public crime reporting on social media as both a progressive and a regressive phenomenon. What does not change within the debate is the understanding that reporting on social media has a larger impact than traditional methods of reporting to offline agencies. Despite the criticisms from scholars, law enforcement and the wider public, social media will continue to represent a coherent space for individuals to become the producers of their own content and share crime experiences. This exploration in this chapter so far suggests that—as the bridges between the online and offline borders continue to dissolve—the debates surrounding public crime reporting will only continue to expand given new technological developments.

Law Enforcement Use of Public Crime Reporting

The penultimate section of this chapter discusses how the data produced by public crime reporting on social media may be gathered and used by law enforcement agencies to aid in investigations. Bousequet (2018) argues that the growing amount of crime reports uploaded online offers a considerable opportunity for law enforcement agencies to improve the effectiveness of investigations. Framing intelligence gathering around the imagery created by public crime reporting can be seen by several law enforcement agencies on an international scale. The contemporary methods used show how law enforcement agencies are combining smartphone applications with public crime reports to encourage the transition back to law enforcement agencies as a place to report crime with the most positive impact. What they particularly focus on is capturing the evidence provided by witnesses of crime to be used alongside their testimony to improve the reliability and accuracy of the witness's perspective (Bousequet 2018).

Paterson (2018) suggests that the aim of this approach is to fill a widening gap in the legal system by allowing witnesses to record and store data that can be used in the courtroom. This indicates a clear focus on protecting the witness's accountability. The approach can be detected in a number of applications developed by law enforcement agencies. For instance, In Nice, French law enforcement agencies have incorporated an application labelled *Reporty*, which takes advantage of the public turning to their phones as the initial response to a crime incident (The Connexion 2018). *Reporty* allows members of the public to livestream, film and report crimes to law enforcement. Police can use the uploaded content to extract video, audio and location of the event. Similar apps have been developed in the UK labelled the *Self Evident App* and in Sydney, Australia, as the *iWitnessed* app, which allow the public to report and capture key real-time evidence following a crime incident that becomes accessible to law enforcement to use in investigations (Kayhan 2018; Stanley 2017; Witness Confident 2019). Witness Confident (2019) highlights the benefit of using innovative tools and their ability to reduce the amount of police time and resources by taking the opportunity of using public crime reports to improve the intelligence breadth of investigations.

What becomes noticeable through researching these nuanced methods is law enforcement's solicitation of remoulding the bystander effect by innovative technological means encouraging the transition of the public from passive spectators of crime to active witnesses, recognising their potential in aiding investigations. Law enforcement is facilitating this conversion by replacing the role of social media as the main platform where crime reports are shared. These examples express an intriguing development in understanding public crime reporting as a phenomenon that provides opportunities for law enforcement. They further display growing law enforcement efforts to capitalise on changing crime reporting behaviours to benefit intelligence gathering and investigations.

These developments have been met with some concern by scholars, who raise the question of unethical practice. Bousequet (2018), for instance, argues that the methods used by law enforcement—such as the gathering and storing of publicly generated data that includes details of the producer, victim and offender—raises concerns regarding public privacy. To overcome this potential obstacle, he argues that agencies should apply a balanced approach to using public crime reporting media. It should be used, as the information is made publicly available by the producer, but agencies should respect that the public desires a reasonable amount of privacy.

A point that should also be deliberated is that despite the efforts and resources that law enforcement agencies invest into these applications they will not be successful without regaining the trust between institutions and the public. As shown in the case examples, many individuals choose to upload crime reports on social media for its ability to enforce social or political impacts. Therefore, future applications will need to demonstrate that they too can have a progressive impact. To achieve these changes, police should aim to demonstrate to the public the value of using publicly produced reports and to show how their use has benefitted wider society; for example, through successful cases that used publicly generated crime reports to prosecute offenders. Such an attempt can be seen, for instance, in the description of the *Self Evident App*, which recognises these growing tensions and highlights how the data uploaded by the public is recognised by the UK Crown Prosecution Service as legitimate evidence, which has secured convictions in the courtroom (Witness Confident 2019). These factors will need to be considered in future developments of law enforcement applications to ensure adequate public—agency cooperation helping to support contemporary investigations. Exploration into the innovative methods used by law enforcement agencies reveals an institutional development that is redefining the phenomenon of public crime reporting.

Discussion

This chapter has aimed to explore public crime reporting on social media as an emerging cultural phenomenon that has become the topic of academic, public and political interest. The nuanced common practice has been examined in detail to highlight the theoretical, interactional and symbolic frameworks that are interwoven in current debates. Taking a cultural criminological approach revealed the symbolically rich nature of public crime reporting, with the goal to provide innovative contributions towards understanding contemporary visual culture (Sandberg and Ugelvik 2017). The chapter has aimed to showcase the phenomenon as a contemporary opportunity for law enforcement agencies to use publicly generated crime reports to improve the effectiveness of intelligence gathering and to rebuild the policing platform as the main space that crime reports are reported to. To fully explore the phenomenon the chapter has applied cultural criminological theories to showcase the impact it has in contemporary society in two ways. The first highlighted the dynamic perspectives and identities that frame how the reports are produced; the latter examined how these features reflect a broad debate into whether the phenomenon represents just behaviour.

Conducting an analysis into the multiple perspectives on crime reporting reveals themes of gender, victimology and social stigmas, influencing the cultural understandings of public crime reporting online. The effects of the incorporation of the camera lens into smart phones shows how technology has allowed the public to produce reports through innovative perspectives and reflects a nuanced set of theoretical frameworks to understand victim and offender identities. The chapter highlights how social activism remains an overarching motivation for crime reporting on social media, which shapes the context of reporting crimes to an interactive online audience. By delving further into the perspective of the offender, the chapter further identifies a new pattern in crime reporting by violent extremists and terrorists producing their own crime reports as a form of propaganda, manipulating social media sites to disseminate their content. This new wave of crime reporting presents an unprecedented path of exploration, which highlights the negative implications of the phenomenon and showcases how the motivations behind crime reporting can be malicious.

What emerges from the comparison of perspectives and identities intertwined within the phenomenon is an overarching debate whether the motivations of public crime reporting are just. This chapter explored this debate through the application of two contemporary case studies. The first case explored the rise of sexual abuse and harassment victims at universities reporting crimes on social media, providing a space for victims to raise awareness and gain support within subcultures. Despite the potential benefits of this practice, the phenomenon has also been labelled as an expression of a regressive society. Adopting the traditional concept of the bystander effect and applying it to online crime reporting reveals a cultural acceleration into a spectator society that has become socially invested in producing visual crime imagery. Overall, current debates reveal how public crime reporting on social media has become a critically examined behaviour within cultural criminology, that reveals wider discussions into the motivations and themes surrounding the phenomenon.

The chapter also offers an analysis into how law enforcement agencies have begun to develop smart phone applications that are redefining how the public should respond to crime situations from a police perspective. The examples discussed reveal that law enforcement agencies are recognising the benefits of using public crime reports to improve the intelligence gathering of witness statements and overall investigations. The methods and applications being developed by law enforcement agencies reflect a cultural transition into a contemporary era of crime reporting, which shows promise for future crime response. The exploration into the phenomenon of crime reporting on social media offers a set of innovative frameworks that contribute towards a contemporary understanding of how individuals respond to crime situations and how crimes are reported within a dynamic technological society.

References

Antony, M. G., & Thomas, R. J. (2010). 'This is citizen journalism at its finest': YouTube and the public sphere in the Oscar Grant shooting incident. *New Media and Society, 12*(8), 1280–1296.

Badalge, K. N. (2017). Our phones make us feel like social-media activists, but they're actually turning us into bystanders. [online]. *Quartz*. [Viewed 1 July 2019]. Retrieved from https://qz.com/991167/our-phones-make-us-feel-like-social-media-activists-but-theyre-actually-turning-us-into-bystanders/

BBC News. (2017). *BBC News*. [Viewed 25 June 2019]. Retrieved from https://www.bbc.co.uk/news/av/uk-39847612/domestic-abuse-victim-who-took-selfies-of-injuries

BBC News. (2019). Facebook: New Zealand attack video viewed 4,000 times. [online]. *BBC News*. [Viewed 5 July 2019]. Retrieved from https://www.bbc.co.uk/news/business-47620519

Becker, H. S. (2008). *Outsiders*. New York: Simon and Schuster.

Black Lives Matter. (2019). *Black Lives Matter*. [online]. Retrieved from https://blacklivesmatter.com/

Bock, M. A. (2016). Film the police! Cop-watching and its embodied narratives. *Journal of Communication, 66*(1), 13–34.

Botsford, P. (2017). Make in unlawful to stand by and record emergency scenes, argues law professor. [online]. *Legal Cheek*. [Viewed 1 July 2019]. Retrieved from https://www.legalcheek.com/2017/06/make-it-unlawful-to-stand-by-and-record-emergency-scenes-argues-law-professor/

Bousequet, C. R. (2018). Why police should monitor social media to prevent crime. [online]. *Wired*. [Viewed 3 July 2019]. Retrieved from https://www.wired.com/story/why-police-should-monitor-social-media-to-prevent-crime/

Busby, E. (2019). More than half of university students experience unwanted sexual behaviour, survey finds. [online]. *The Independent*. [Viewed 1 July 2019]. Retrieved from https://www.independent.co.uk/news/education/education-news/university-students-sexual-harassment-behaviour-campus-brook-uk-a8796531.html

Darley, J. M., & Latané, B. (1968). Bystander intervention in emergencies. *Journal of Personality and Social Psychology, 8*(4), 377–383.

Domestic Shelters. (2014). Survivor Selfies: Does posting online photos of the aftermath of abuse help or hurt the cause? [online]. *Domestic Shelters*. [Viewed 5 July 2019]. Retrieved from https://www.domesticshelters.org/articles/ending-domestic-violence/survivorselfies

Eler, A. (2017). *The selfie generation: How our self images are changing our notions of privacy, sex, consent and culture*. New York: Simon and Schuster.

Greer, C. (2006). News media, victims and crime. In P. Davies, P. Francis, & C. Greer (Eds.), *Victims, crime and society* (pp. 20–50). London: Sage.

Hill, K. (2016). 'Gawker' behaviour at fire and accident scenes getting in way of emergency responders. [online]. *ABC News*. [Viewed 1 July 2019]. Retrieved fromhttps://www.abc.net.au/news/2016-01-28/gawkers-asked-to-reconsider-behaviour-by-emergency-services/7121792

It Happens Here. (2019). *It happens here*. [Viewed 1 July 2019]. Retrieved from https://ithappenshere-oxford.tumblr.com/about

Kayhan, I. (2018). Phone app helps witnesses record evidence. [online]. *SBS*. [Viewed 3 July 2019]. Retrieved from https://www.sbs.com.au/yourlanguage/turkish/en/article/2018/04/03/phone-app-helps-witnesses-record-evidence

Lewis, R., Marine, S., & Kenney, K. (2018). I get together with my friends and try to change it. Young feminist students resist 'laddism,' 'rape culture' and 'everyday sexism'. *Journal of Gender Studies, 27*(1), 56–72.

Lombardo, C. (2019). 5 arrested after homophobic attack on London bus. [online]. *NPR*. [Viewed 22 June 2019]. Retrieved from https://www.npr.org/2019/06/08/730951943/five-teens-arrested-in-homophobic-attack-on-london-bus?t=1561641184906

Miller, V. (2011). *Understanding digital culture*. London: Sage.

Mills, S. (2003). *Michel Foucault*. Canada: Routledge.

Office for National Statistics. (2019). How much do you really know about crime? [online]. *Office for National Statistics*. [Viewed 1 July 2019]. Retrieved from https://www.ons.gov.uk/peoplepopulationandcommunity/crimeandjustice/articles/howmuchdoyoureallyknowaboutcrime/2018-11-06

Oppenheim, M. (2019). Students share sexual abuse allegations online because they feel ignored by university staff. [online]. *The Independent*. [Viewed 1 July 2019]. Retrieved from https://www.independent.co.uk/news/uk/home-news/student-sexual-abuse-assault-harassment-university-social-media-a8927201.html

Pande, S. (2017). Ethics in citizen journalism: Incident of teenage girl molestation in India. *Journal of Information, Communication and Ethics in Society, 15*(1), 2–16.

Paterson, H. M. (2018). Why we made iWitnessed, an app to collect evidence. [online]. *The Conversation*. [Viewed 3 July 2019]. Retrieved from https://theconversation.com/why-we-made-iwitnessed-an-app-to-collect-evidence-94107

Sandberg, S., & Ugelvik, T. (2017). Why do offenders tape their crime? Crime and punishment in the age of the selfie. *The British Journal of Criminology, 57*(5), 1023–1040.

Solon, O. (2017). Why a rising number of criminals are using Facebook live to film their acts. [online]. *The Guardian*. [Viewed 1 July 2019]. Retrieved from https://www.theguardian.com/technology/2017/jan/27/rising-numbers-of-criminals-are-using-facebook-to-document-their-crimes

Somerville, E. (2019). Sheffield University student investigated over offensive messages about women and STIs in leaked group chat. [online]. *The Independent*. [Viewed 1 July 2019]. Retrieved from https://www.independent.co.uk/news/education/education-news/sheffield-university-student-derogatory-messages-women-chat-group-stis-a8771286.html

Stanley, J. (2014). Private cameras will hurt privacy—But is there a solution? [online]. *ACLU*. [Viewed 27 June 2019]. Retrieved from https://www.aclu.org/blog/national-security/privacy-and-surveillance/private-cameras-will-hurt-privacy-there-solution?redirect=blog/national-security-free-speech-technology-and-liberty/private-cameras-will-hurt-privacy-there

Stanley, J. (2017). A new implication of cellphone video: Citizens taping each other for police. [online]. *ACLU*. [Viewed 27 June 2019]. Retrieved from https://www.aclu.org/blog/privacy-technology/surveillance-technologies/new-implication-cellphone-video-citizens-taping

Stokes, J. (2013). *How to do media and cultural studies.* London: Sage.

Tatham, H. (2016). Breaking the 'bystander effect' and saving lives. [online]. *ABC News.* [Viewed 1 July 2019]. Retrieved from https://www.abc.net.au/news/2016-04-05/breaking-the-bystander-effect-and-saving-lives/7300152

The Connexion. (2018). Nice locals asked to film crimes in new police app. [online]. *The Connexion.* [Viewed 3 July 2019]. Retrieved from https://www.connexion-france.com/French-news/Nice-locals-asked-to-film-crimes-in-new-police-app

Uelmen, A. J. (2017). Crime spectators and the tort of objectification. *University of Massachusetts Law Review, 12*(1), 68–123.

Witness Confident. (2019). *Witness confident.* [online]. Retrieved from https://witnessconfident.org/

Yamasaki, A. (2019). #StandUpJapan and other hashtags allow for new voices to be heard. [online]. *Japan Times.* [Viewed 1 July 2019]. Retrieved from https://www.japantimes.co.jp/community/2019/01/17/general/standup-japan-hashtags-allow-new-voices-heard/#.XRoLmflKhhE

Yar, M. (2012). E-Crime 2.0: The criminological landscape of new social media. *Information & Communications Technology Law, 21*(3), 207–219.

CHAPTER 6

Community Policing: A Case Study from Bavaria Within the Framework of the Unity Project

Holger Nitsch and Sebastian Allertseder

Introduction

During 2015, a higher number of refugees came to Europe. Between 800.000 and 1.1 million (BiB 2017)—480.000 from Syria, Afghanistan, Pakistan and Iraq (BAMF 2016, p. 2)—were seeking for asylum in Germany. Most of them came through Bavaria due to its location, and many of those who aimed to continue their journey to other countries north or west were passing through the region of Bavaria. There were fears that in some regions this immigration process could affect changes in the social structure. Besides the impact of immigration on the social-cultural structure especially in smaller towns, fears were raised that the refugees could have an effect on the overall security situation and the way of living of German citizens. For the police this situation of growing fears

H. Nitsch (✉) · S. Allertseder
University of Applied Sciences for Public Administration,
Fuerstenfeldbruck, Bavaria, Germany
e-mail: holger.nitsch@pol.hfoed.bayern.de;
sebastian.allertseder@pol.hfoed.bayern.de

B. Akhgar et al. (eds.), *Social Media Strategy in Policing*,
Security Informatics and Law Enforcement,
https://doi.org/10.1007/978-3-030-22002-0_6

and a population divided by the questions of immigration, was and still is one of the biggest challenges of the last decade. There was a balancing act between the real mission of the police to keep up public order and security, on the one hand, and a partly paranoid xenophobic atmosphere among some groups, which had no connection to the objective crime statistic, on the other hand. This imbalance showed the necessity of a preventive, reliable and widely recognized communication culture between police and citizens in order to face misinformation and untruthfulness distributed in the Internet and social media.

In addition, most of the refugees had made their own experiences with the police in their home countries, which often differ markedly from the situation in Germany. Therefore, on the one hand, the police had to deal with the fears and anxieties of the citizens, and on the other hand, they were confronted with people whose experiences with the police usually were not very positive and very often determined by corruption and violence. Anyhow, mistrust and uncertainty marked the inactions between both groups.

Furthermore, at the same time European countries were hit by a growing number of attacks by terrorist groups with an Islamist background. In the aftermath the tensions against Islam and Muslim communities were also rising. The opponents of the migration politics of chancellor Merkel and her government felt themselves confirmed in their critical view, especially against immigration of Muslims.

Therefore, there was a need of a new strategic way to challenge these issues. As media use is no longer only focused on traditional media, an intensification of the use of social media channels was unavoidable. Because not all information was transported in a correct way through these channels, an intensified use of social media through the police has not only the aim to keep followers informed, but especially to set a strong counterpoint against wrong information and "alternative facts".

As the aim for uniformed policing was to be close to the citizens, the same proximity had to be created to the new groups inside the community. A side effect of this new strategy would be to get also closer to the majority of the people, but also to other minority communities, which are not using traditional communication channels. This increase of social media use through the police is considered to be a part of the community policing strategy. The examples used in this chapter will concentrate on this issue by analyzing this strategy during a real case.

Community Policing in Bavaria

Although the expression community policing (CP) is not officially used by the Bavarian Police, a German synonym would be "bürgernahe Polizei", which stands for policing close to the citizens and means basically the same.

Community policing (CP) is part of policing strategies and practices. It can be seen as a strategic tool or policing strategy to improve the prevention as well as the investigation of crimes (Nitsch and Ronert 2017) with the aim of strengthening the cooperation between communities and the police organization. This was also the goal of the EU funded H2020 project Unity. According to the results of the research within the Unity-project (2017), the key elements for successful CP are (Hail and O'Neill 2016; Unity Brochure 2017, p. 10; van der Giessen et al. 2016):

- Trust and confidence building.
- Accountability.
- Information sharing and communication.
- Addressing local needs.
- Working in a collaborative manner.
- Crime prevention.

As there is no uniform concept of CP, it is more an interaction of policing and communities orientated at the needs and the structure of the region or the existing communities. The conversion within the Bavarian Police started in the 1980s to change policing from a top-down approach to a multilevel approach amongst different actors. This change was accompanied by apprehensions and biases by some, because critics feared the loss of traditional policing and a lack of securing the society. However, in fact this "soft on crime" shift (Ferrandino 2013, p. 64) did not take place. Instead a number of new policies were introduced to support the above-mentioned approach by building new partnerships and the principle of decentralization as described by Skogan (1990). Schrödl (2017) elaborated in his thesis on CP in the context of migration communities in a certain region in Bavaria. He combined, as a basis for successful CP, the *Broken Windows Theory*, the *Problem-Orientated Approach* of Goldstein and the *Neighbourhood-Cohesion Theory* of Sampson (Schrödl 2017), which is, besides the Bavarian example, also applicable for many other regions and cities. The result of his work reflects very well the results of the Unity research. The most relevant elements of successful CP are trust, accountability, working collaborative and crime prevention. As both studies were conducted independently, it is most likely that the results are relevant factors for a successful CP in Bavaria.

A very special focus has to be set on the development of media and the change of their use by the people in recent years. Due to the introduction of social media and Web 2.0, the role of media in CP increased dramatically. Opinion building is now in some groups of the society very much focused on the information gained through social media. This is why the provision of reliable and true information plays an important role and is relevant for an open and democratic society.

Especially social media, with its quick and far-reaching transfer of information, became an important factor for strengthening the communication between the police and the communities (cp. Chap. 1). For the police the use of social media offers the possibility for nearly unlimited information sharing, which is especially needed during crises (Schmitt 2014, p. 44). Because of this, many law enforcement agencies (LEAs) have already recognized the opportunities of social media and started to use the Internet and social media to spread information as well as to improve their image and to communicate with citizens. Overall, CP means to change the role of LEAs from a static, reactive, incident-driven bureaucracy to a more dynamic, open, quality-oriented partnership with the community (Brown and Wycoff 1987). Common elements within different community-oriented approaches are the exchange of trust, an increased officer-accountability to civilians and the fostering of trust. This can be achieved by also using social media for presenting the human face of police (cp. Chap. 4). By using a more confidential style of writing the trust between users and police can be strengthened by showing the police as a service provider for the citizens and not only as a bureaucratic agency not related to the daily life of people (Hirschmann 2013, p. 4). In the long term these police–citizen partnerships will also strengthen cooperation and will improve the prevention and the investigation of crimes, increase the subjective feeling of safety and reduce the risk of radicalization (OSCE 2014).

Especially in times of the refugee crisis and social changes, these aims are more important than ever. LEAs regularly face different forms of radicalization, which are arising in all kinds of social classes, age groups, nationalities and political groups.

Main Objectives of the Bavarian Pilot

At the end of 2015, a pilot was conducted at the University of the Bavarian Police within the framework of the Unity project. At this stage, surveys about user requirements and stakeholder analysis had already been conducted and the architectural framework of the project was finished and adjusted to the pilot. Because of the high number of refugees entering Germany mainly via Bavaria, the community chosen were refugees and NGOs dealing with and helping these refugees. In addition, there was a need to invite other relevant stakeholders to get their views on the innovation of the project. These were mainly from police forces.

While discussing the disparate needs of the invited communities in small groups, the refugees talked about their experiences and needs during the immigration process. Best practice models like the Finnish one (cp. Chap. 4), where Iraqi refugees received smart phones and apps to share information and videos about what they need to know (who to contact, what the weather is like, what clothing they will need, etc.), were considered. However, it was also recognized that every use of technical solutions is limited, because of previous negative experiences of many refugees with police, an important percentage of them is not willing to use technology provided by the police, in order to avoid surveillance and gathering of information through state actors. This attitude is not only problematic within the refugee group, but also a widespread attitude among many citizens. Finally, the role of the police was described as a link between refugees and local community. The use of modern forms of technology is part of the digital community policing strategy (Bayerl and Rüdiger 2017). In case of rumours about refugees affecting local areas, police can use social media to publish stories advising the rumours are incorrect and deescalate the situation.

In contrast to this, the means of communication between the refugees/migrants was seen very positively, and the possibility to interact with the partners (NGOs/police) was considered as a big opportunity for all parties. It seemed that the content of the information is the most relevant factor for the use of the technical application. The design of the tool and the tailor-made content relevant to the minority group are also key factors for a successful acceptance of the technical innovation. As the design is foreseen to have a local focus and relevant information for the user living in the region—for instance, the possibility to see announcements for events in a different place, where friends live—was also considered to be a step to build trust between the different groups. Building trust is relevant as many migrants have friends and relatives also in other areas not just in Germany, but also in Europe. By that, they would be able to communicate with their friends via the app. In addition, NGOs would like to share information between the partners and have access to a proven and trusted information source. Especially the last point was very important for NGOs present, after experiences of frequently dealing with disparate and politically exploited information about the same event. This also corresponds with Perkin's work (2013), who describes the importance of providing online services to engage communities online.

As CP is usually meant to influence the day-to-day service of police officers and to build a relationship with the communities, nowadays the need to use social media is evident and crucial for the success of CP. It is not limited to the daily routine; it can also be used during emergencies. It has to be pointed out that CP should be conducted as a normal part of the daily work of an officer or an LEA organization, but in an emergency as described next, the positive effect might be seen even clearer, because of the short time-span and the speed that information is shared among the community. In addition, the important role of social media use is evident with all its advantages, but also with its challenges in an emergency situation.

The Incident

On the 21st of July 2015, a young man posted on the Facebook account of a young girl he had hacked previously that he would like to meet people at 4 pm in a McDonalds Restaurant close to a shopping Mall in the north of Munich called the "Olympia Einkaufszentrum" (OEZ). In this post, he wrote that he would be willing to pay something for the ones coming, as long as it was not too expensive. In the early evening of the 22nd, he entered the fast food restaurant and started shooting, especially on younger adults with a visible migrant background. He continued to do so outside the restaurant. Then he ran into the shopping mall, started firing and shouted: "I don't care, I kill you all" (Spiegel Online 2016). The offender David S. ran off to a nearby car park, where he was filmed on the upper deck, and communicated with an inhabitant of a house on the opposite side of the street, saying that he was German, bought a gun and is in therapy.

Officers entered the park deck and shots were fired, but David S. was able to escape. At 8:30 pm, he was seen by a police patrol in a park. Seconds after the first contact with these officers, he pulled his gun and shot himself in the head. At the end of this rampage, altogether nine people were dead and of those eight were aged between 14 and 20 and mostly of Turkish and Kosovo origin. Four others were injured and treated in hospital. In addition, a higher number of people injured themselves during the panic in the shopping mall and at different places in the city.

During the shooting, next to the traditional media Twitter played a crucial role in informing citizens about the situation. A young man reporting at the scene sent out the first live coverage via social media. In very short time, this post had gained more than 100.000 followers. The Munich Police Department had around 700 calls per hour—four times more than usual. Additionally, the posts on Twitter related to the event added up to approximately 113.000, not including the extremely high number of posts and messages sent by other social media channels like Facebook, Snapchat, Instagram and WhatsApp (Backes et al. 2016). Shortly before the shooting, there was an attack in Nice, France, where a terrorist killed more than 80 people with a truck. Therefore, there was a mix of words being used during the offence between the expressions "terror" and "amok". "Terror" makes citizens more fearful. In an analysis of the Twitter feeds, the word "amok" was used 2978 times and "terror" 58.237 times (ibid). At 7:20 pm the user *@itsflyingbird* posted that there are also shots heard at the Stachus, which is another central place in the inner city about 8 km away from the real shooting. Despite the fact that this user had just 160 followers, this message was mentioned 1600 times shortly after and about 58 times per minute. In addition, a nearby reporter at the OEZ read it and mentioned it in a phone interview with a nationwide TV-station. Therefore, the phrase "shooting" was quickly changed to "heavy shooting" and rumours spread that the offenders are running towards the central train station, which is also located about 7 km away from the original crime scene. Because of this, many people spread through social media that the next crime scene will be or already was the central train station (ibid.). The distance of these different crime scenes made it very likely, from the authors' point of view that there had to be more than one perpetrator involved, what lead to rising panic within the city.

A journalist entering a taxi at the central station saw a message on the screen of the car to avoid taking passengers into the inner city and to avoid the inner city. She took a photo and sent it via Twitter. It was distributed by 400 followers and read at least one million times. As a result, public transport into the inner city was shut down completely. At the traditional restaurant Hofbräuhaus—a major tourist attraction in Munich—at about 7.15 pm a man ran into the building shouting "shootings, shootings" and suddenly all people tried to get out in panic and fear. One woman from Macedonia jumped out of a window and was seriously injured. This was also reported in a video and posted on Twitter (ibid). In contrast to this behaviour, the visitors of a festival nearby left the area quietly and without any incidents.

The comments on social media throughout this event were surprisingly realistic, and all of the comments about additional shootings received denials. A safety check similar to the incident in Nice was activated on Facebook, and many citizens started to offer shelter under the hashtag *#opendoors*. Besides these bottom-up phenomena, also the tweets of the Munich police were shared very often, already demonstrating the interest of people to get first-hand information about the situation. One tweet requesting people to stay away from the streets was retweeted 7.000 times, and the plea not to upload any videos was retweeted even more often—47.000 times (ibid). In total, there were 36 injuries related to the event including the real victims of David S., but also people at faraway places that injured themselves by seeking shelter during the panic. Even after the offender killed himself (8:29 pm), there were still reports on Twitter about shootings or shots being heard in the city. At 8:38 pm, directly after the first report about the death of at least one of the possible perpetrators, there was the record number of 366 tweets per minute about the situation (ibid). The activity on Twitter continued until 4 am the next day, and the last climax was around 1:30 am when people started retweeting a post of the police saying that the danger is over.

Analysis

Taking the effect of social media into account, it is obvious that from the perspective of an LEA an active and well-thought through participation in the process of generating on-time news is necessary to calm down the situation, avoid further panic and to counter misinformation by gaining the trust of the citizens. Therefore, it is necessary to analyze the tweets of the police on the one side and on the other side to have a closer look to the retweets and the hashtag *#OEZ*. To do this the authors used the Unity analytics tool, designed by the project consortium.

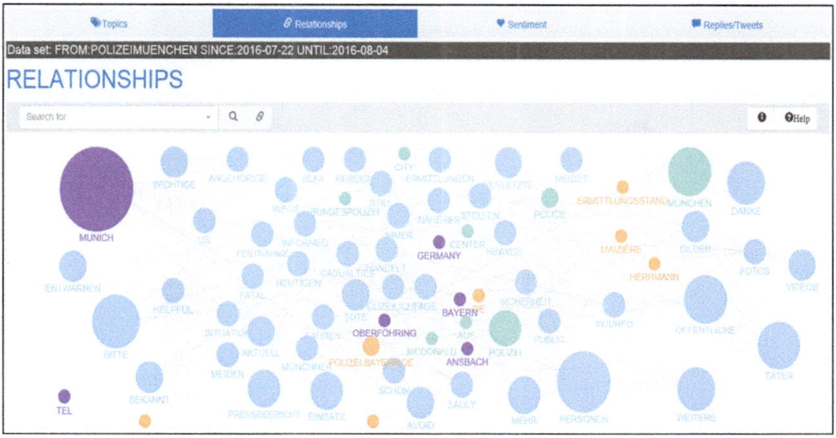

Fig. 6.1 Relationship between the words used in tweets from Munich Police Department (Source: Unity analysis)

The most frequently used word in the tweets of the Munich Police sent during and shortly after the event, not surprisingly were focussing on the location and the incident. Of course, words like "offender" and "victims" ("Täter", "Opfer") were also used very regularly, but not to the same degree as the location. As can be seen in Fig. 6.1, some of the posts were also in English. Due to the fact that Munich has a very multicultural population and high number of foreign communities, this measure expands the reach of the tweets and is taking the needs of the different communities into account. It can be seen that also another kind of word was very dominant: *Please* ("Bitte") and *thank you* ("Danke"). The word *please* was used more often and was always connected with a request to citizens. Most of the words have a neutral connotation. Due to the situation it is unavoidable to use negative phrases, but the public relations department of the Munich Police tried to use phrases that were as neutral as possible.

The word "terror" was not used, which shows that the wording was meant to calm down the situation and to be as close to the trusted facts as possible. In addition, the names of the Minister of the Interior of Germany, De Maiziere, and the Minister of Interior of Bavaria, Herrmann, were mentioned, to show that there is a concern and action is being taken on a high political level. Figure 6.2 shows examples of frequently retweeted posts during the event.

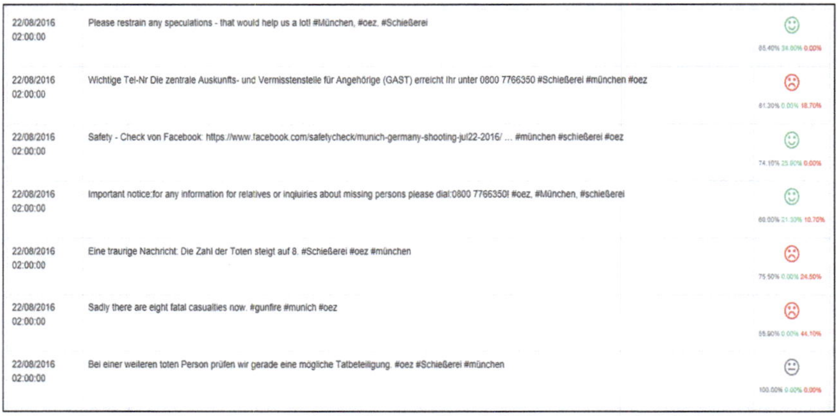

22/08/2016 02:00:00	Please restrain any speculations - that would help us a lot! #München, #oez, #Schießerei	65.40% 34.60% 0.00%
22/08/2016 02:00:00	Wichtige Tel-Nr Die zentrale Auskunfts- und Vermisstenstelle für Angehörige (GAST) erreicht Ihr unter 0800 7766350 #Schießerei #münchen #oez	81.30% 0.00% 18.70%
22/08/2016 02:00:00	Safety - Check von Facebook: https://www.facebook.com/safetycheck/munich-germany-shooting-jul22-2016/ ... #münchen #schießerei #oez	74.10% 25.90% 0.00%
22/08/2016 02:00:00	Important notice:for any information for relatives or inqiuiries about missing persons please dial:0800 7766350! #oez, #München, #schießerei	68.00% 21.30% 10.70%
22/08/2016 02:00:00	Eine traurige Nachricht: Die Zahl der Toten steigt auf 8. #Schießerei #oez #münchen	75.50% 0.00% 24.50%
22/08/2016 02:00:00	Sadly there are eight fatal casualties now. #gunfire #munich #oez	55.90% 0.00% 44.10%
22/08/2016 02:00:00	Bei einer weiteren toten Person prüfen wir gerade eine mögliche Tatbeteiligung. #oez #Schießerei #münchen	100.00% 0.00% 0.00%

Fig. 6.2 Most frequently retweeted posts originating from the Munich Police Department (Source: Unity analysis)

It shows that the public relations department of the Munich Police tried to communicate in several languages. Besides English there were also French, Spanish, Italian, Croatian and some other languages. One post requesting people not to upload any pictures or videos was repeated very often and translated in order to guarantee a maximum level of distribution throughout the channels. Also, the request to stay at home or at the current location and not to move outside onto the street was constantly repeated. There was also a reference on the Facebook site of the police department for further information. By this the department tried to get the lead in giving information and trying to counter false information, which at this time spread as described in the previous section.

Concerning the posts tweeted using the hashtag *#OEZ* (Fig. 6.3), it is obvious that the most frequently used word was *thank you* ("Danke").

It can also be clearly seen that the tweets of the Munich police department were frequently mentioned and retweeted, which gives a clear hint about the success of their efforts to claim the leading role in sharing information about the situation. Reading the posts, it is obvious that the connotations are very positive, which leads to the assumption that the work of the public relations department of the police was very much appreciated. The words "Pressesprecher" (PR manager), "Pressekonferenz" (press conference) and "Martins" (the name of the PR manager of the Munich police department), are mentioned quite often. In addition, the work of

Fig. 6.3 Most frequently used words in Hashtags (#OEZ) (Source: Unity analysis)

the police was often seen as good, which is demonstrated by the high number of postings containing the phrase *good job* ("guter Job/gute Arbeit").

The fact that many posts were in English led also to a high number of responses in English. This means that the posts in English were actually read and that there was a high number of none-German-speaking individuals who interacted with the police or at least used their social media channels as a preferred source of information.

For example, as shown in the selected posts in Fig. 6.4, many retweets just repeated the information given by the police or repeated the requests of the police without any personal comments. The second tweet from the bottom refers to the request not to upload videos and pictures. This person asked instead for the upload of pictures of cats, which many followed, and hundreds of pictures of cats were uploaded. Others asked for reading the tweets of the police to get the relevant information. Also, some politically incorrect posts can be found, although their number was rather low and did not play a major role amongst the big amount of data sent out minute by minute. The total number was so low, that these postings and the used hate speech cannot even be seen in the word analysis.

The sentiment analysis (Fig. 6.5) illustrates mostly neutral content, or according to the event negative content, although it can be seen that also positive tweets were posted. Looking into these in more detail, they were mostly about the work of the police and the public relations management.

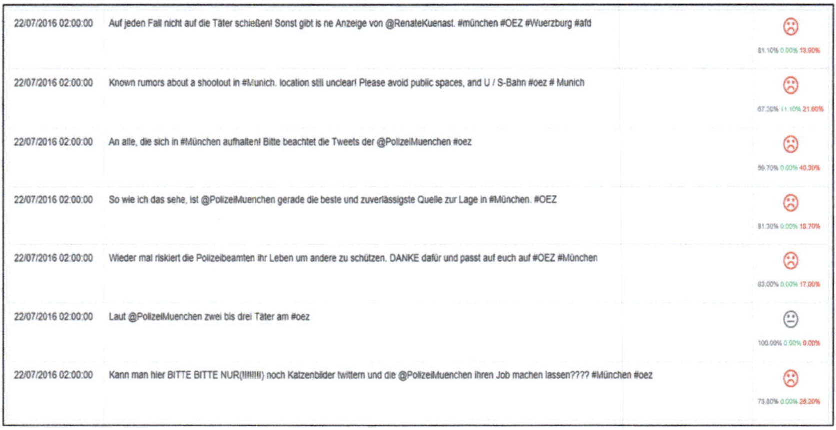

Fig. 6.4 Examples of retweeted warnings and information from Munich police department (Source: Unity Analysis)

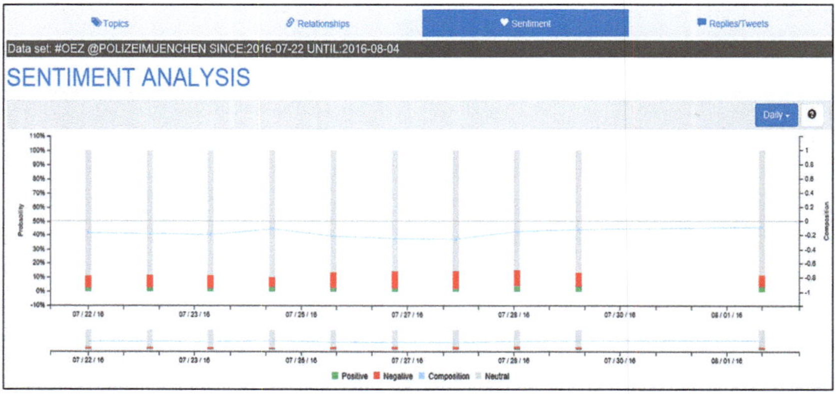

Fig. 6.5 Sentiment Analysis of the tweets (Source: Unity Analysis)

It seems that most people, even after some misinformation and rumours spread, intended to believe the information issued by the police department, and even furthermore that they started repeating the neutral wording.

Conclusion

As mentioned at the beginning of the chapter, CP should not exclusively work for daily police services for and with citizens and communities, but also in emergency or crisis situations. An example as the one above shows what CP can do to gain trust and accountability by addressing the local needs in a proper way. The role of social media is evident in such a situation, and CP has to play its role. As communication in social media is very fast, the job for officers dealing with these channels will not be easy. As the research by Fichet et al. (2016) shows, communication via social media by governmental institutions in a crisis can influence the public opinion positively, especially regarding the fact that Twitter and Facebook are substitutes for personal eye witnessing. The correct use of social media and especially Twitter revitalizes conversations and also counters misinformation. This can also encourage users that posted false information to correct themselves (ibid). During a crisis situation, more than in daily routines, people are searching for reliable and direct information of an incident. If there is a lack of trusted information sources, which is usually the case (for instance, at the beginning of the crisis) rumours will start to fill this gap. Therefore, it is necessary to react accordingly and to respond quickly. It is a necessity for modern police work to mark this kind of uncertain information and to provide instead trusted information. In the reality of this crisis we used as an example, there was just one crime scene, but according to the discussions in social media, the number added up to 67 locations. By constantly posting the state of the investigations and showing that the police works as transparently as possible, most of these cases were quickly uncovered as false alarms and rumours were not retweeted anymore. Of course, this means additional work for police forces, because they have to react to the information and the rumours in a highly stressful situation; but the more people rely on the information given by governmental officials, the easier and the quicker false information can be corrected. At the end of this process, critical situations can be controlled and solved faster. This has a clear link to the aims and goals of the CP concept.

The analysis has its weaknesses concerning the elaboration of words, but it proves that the use of social media has positive effects. The situation—as bad as it might be—can be calmed down and true and correct information can be sent to the public.

Nevertheless, this example from an extreme crisis situation can be transferred to the daily routine. Sharing information in an appropriate way and communicating in a collaborative manner leads to a strengthening of the trust and the accountability of the LEAs. The Unity analysis tool can be helpful to analyze and self-reflect the own way of communication as well as the responses in order to measure the success of the own work. The multilingual approach of the posts of the Munich Police Department proved to be successful as can be seen by the retweets. This can be transferred to daily routines as well. Different communities, some with difficulties understanding German, will be able to obtain the necessary information, and by doing so, their trust and confidence into police will be strengthened.

Social media is used by most people nowadays and especially by minority communities. This advantage should be considered by sharing information with these communities to gain their trust and confidence.

References

Bayerl, P. S., & Rüdiger, T.-G. (2017). Die polizeiliche Nutzung sozialer Medien in Deutschland: Polizei im digitalen Neuland. In J. Stierle et al. (Eds.), *Handbuch Polizeimanagement. Polizeipolitik—Polizeiwissenschaft—Polizeipraxis* (pp. 919–943). Wiesbaden: Springer Fachmedien.

Backes, T., Jaschensky, W., Langhans, K., Munzinger, H., Witzenberger, B., & Wormer, V. (2016). Timeline der Panik. *Süddeutsche Zeitung*. Retrieved October 1, 2017, from http://gfx.sueddeutsche.de/apps/57eba57891 0a46f716ca829d/www/.

BAMF – Bundesamt für Migration und Flüchtlinge. (2016). *Migrationsbericht 2015—zentrale Ergebnisse. Nürnberg*. Retrieved October 1, 2017, from https://www.bamf.de/SharedDocs/Anlagen/DE/Publikationen/Migrationsberichte/migrationsbericht-2015html.

BiB – Bundesinstitut für Bevölkerungsforschung. (2017). *Räumliche Bevölkerungsbewegung. Wiesbaden*. Retrieved October 1, 2017, from http://www.bib-demografie.de/SharedDocs/Glossareinträge/DE/R/raeumliche_bevoelkerungsbewegung.html.

Brown, L., & Wycoff, M. A. (1987). Policing Houston: Reducing fear and improving services. *Crime and Delinquency, 33*(1), 71–89.

Ferrandino, J. (2013). Community policing—And what it is not. In K. Peak (Ed.), *Encyclopaedia of community policing and problem solving*. Los Angeles: Sage.

Fichet, E., Daily, D., Robinson, J., & Starbird, K. (2016). Eyes on the ground - Emerging practices in periscope use during crisis events. In A. Tapia, P. Antunes, V. A. Bañuls, K. Moore, & J. Porto (Eds.), *ISCRAM 2016 Conference Proceedings—13th International Conference on Information Systems for Crisis Response and Management*. Rio de Janeiro, Brazil: Federal University of Rio de Janeiro. Retrieved October 15, 2017, from https://static1.squarespace.com/static/56adaa71f699bb2b315e23ed/t/56b1a80204426258l6468f1e/1454483459185/ISCRAM2016_Periscope_FINAL.pdf

Hail, Y., & O'Neill, M. (2016). *Unity—Report on the social and cultural limitations of an ICT tool for community policing*. Deliverable 2.2. Retrieved October 14, 2017, from https://www.unity-project.eu/wp-content/uploads/2015/12/D2.2-Report-on-the-Social-and-Cultural-Limitations-of-an-ICT-Tool-for-Community-Policing.pdf.

Hirschmann, N. (2013). Beispiele für die Verwendung Sozialer Medien in den Polizeien Europas. *PVT—Fachzeitschrift für Polizei- und Verkehrsmanagement, Technik und Ausstattung, 6*, 2–4.

Nitsch, H., & Ronert, S. (2017). Community policing and radicalization: Evaluation and European examples. In P. S. Bayerl, R. Karlovic, B. Akhgar, & G. Markarian (Eds.), *Community policing—A European perspective. Strategies, best practices and guidelines*. Cham: Springer.

OSCE. (2014). *Preventing terrorism and countering violent extremism and radicalization that lead to terrorism—A community policing approach*. Retrieved October 1, 2017, from http://www.osce.org/atu/111438?download=true.

Perkins, E. (2013). Website uses by local agencies. In K. Peak (Ed.), *Encyclopedia of community policing and problem solving*. Los Angeles: Sage.

Schmitt, J. (2014). Social Media—Chancen und Risiken von Facebook und Co. im Rahmen der Polizeiarbeit. In R. Berthel (Ed.), *Polizei im digitalen Zeitalter. Teil 1*. Rothenburg: Polizeiwissenschaftliche Schriftenreihe Band 70.

Schrödl, M. (2017). *Bürgernahe Polizeiarbeit im Kontext migrationbedingter erausforderungen*. Masterthesis, DHPol, Münster.

Skogan, W. (1990). Probleme gemeinwesenorientierter Polizeiarbeit am Beispiel des Houston-Projekts. In T. Feltes & E. Rebscher (Eds.), *Polizei und Bevölkerung—Beiträge zum Verhältnis Polizei und Bevölkerung und zur gemeindebezogenen Polizeiarbeit ("Community Policing")*. Holzkirchen, Oberbayern: Felix Verlag.

SPIEGEL Online. (2016). *Kommt ins Meggi am OEZ*. Retrieved October 12, 2017, from http://www.spiegel.de/panorama/justiz/amoklauf-in-muenchen-so-war-der-tathergang-a-1104418.html.

Unity-Project. Retrieved October 12, 2017., from www.unity-project.eu.

Unity Brochure. (2017). Retrieved October 12, 2017, from https://www.unity-project.eu/wp-content/uploads/2015/12/Unity_Brochure.pdf.

Van der Giessen, M., Bayerl, P. S., Jacobs, G., & Brein, E. (2016). *Unity - 1st stakeholder analysis—Comparative view on stakeholder needs and perspective. Deliverable 3.3*. Retrieved October 14, 2017, from https://www.unity-project.eu/wp-content/uploads/2015/12/D3.3-Comparative-View-on-Stakeholder-Needs-and-Perspectives.pdf.

CHAPTER 7

From Vigilantism to Digilantism?

Frauke Reichl

Introduction

In the modern societies of countless nations across the globe, citizen groups are uniting in response to a variety of (perceived) social threats. The members of the *Minutemen Project* in the USA are monitoring the Arizona–Mexico border for the location of border crossers who might be illegal immigrants (Chavez 2005). After demonstrations in London regarding a 29-year-old's death during his detention had ended violently in 2011, people got together to protect their homes and businesses by themselves (Arfsten 2012), while a citizen initiative in Germany acts against crime stemming from the near border of the Czech Republic (Schmidt-Lux 2012).

Project PluS-i: Pluralisation of local urban security production — Funding: The research leading to the results presented in this chapter is funded by the Federal Ministry of Education and Research (BMBF) under the Federal Government's framework programme Research for Civil Security. The project website can be accessed at www.plus-i.de/en

F. Reichl (✉)
Westfälische Wilhelms-Universität Münster, Münster, Germany
e-mail: frauke.reichl@uni-muenster.de

B. Akhgar et al. (eds.), *Social Media Strategy in Policing*,
Security Informatics and Law Enforcement,
https://doi.org/10.1007/978-3-030-22002-0_7

These groups as well as the more spontaneous actions can be described as vigilante groups or vigilantism. *Vigilantism*, which means very broadly citizen groups who want at least ostensibly to provide safety and security often by taking the law into their own hands, is an old phenomenon with a long tradition especially in the USA. However, it appears as a worldwide phenomenon—even in new and different kinds. With the emergence of the World Wide Web and its social media a new kind of vigilantism, *digital vigilantism* (DV) (Trottier 2017) or *digilantism*[1] (Arfsten 2012), can be observed.

Trottier describes DV as "a process where citizens are collectively offended by other citizens' activity, and respond through coordinated retaliation on digital media, including mobile devices and social media platforms" (Trottier 2017, p. 56). As will be shown in the following sections, there is a broad range of online phenomena that appear in the context of digital vigilantism. Thus, the chapter addresses the question whether there is a shift from vigilantism to digilantism or a continuing existence of conventional vigilantism through new means.

Vigilantism: An Ambiguous and Contested Concept

Vigilantism is an emotional, variable and multifaceted phenomenon that is contextual and mostly short-lived (Arfsten 2012). There is little empirical material available as vigilantes—due to the informal or sometimes even illegal nature of vigilantism—often avoid the public sphere and rarely build "long-lasting social institutions" (Kowalewski 2003, p. 340). Therefore, Kowalewski describes vigilante action as "fire fly events"(Kowalewski 2003, p. 340).[2]

Despite the thin research landscape about vigilantism, there are a number of scientists engaged in developing concepts, theories and definitions concerning vigilantism. These concepts and definitions on vigilantism developed by various authors differ, for instance, concerning the inclusion of state actors, force or threatening of force, the description as a social movement, the necessity that vigilantes take the law into their own hands

[1] Digilantism and digital vigilantism (DV) will be used synonymously throughout this article.

[2] Although digital vigilantism has more written empirical material available, Trottier underlines that the material is sometimes only available for a short period of time as social platforms remove it "due to legal, ethical and moral reasons" (Trottier 2017, p. 58).

and the lost trust in the effectiveness of the state by vigilantes.[3] All in all, it can be stated that a large part of the available research on vigilantism explains its emergence with fighting off disliked behaviour or with the preservation of own values and norms, respectively. Additionally, there is an inherent tension with the state's monopoly of force. Overall, vigilantism is a highly problematic and contested field: it often works in a legal and moral grey area (Arfsten 2012). Vigilantism can be very broadly seen as citizens or citizen groups that want at least ostensibly provide safety by fighting disliked or criminal behaviour and in this process often take the law into their own hands. Thereby, they aim at the preservation of their values and norms.

In this chapter, only two broad theoretical angles are considered. The other theoretical approaches are not taken into account, because on the one hand the focus is on empirical examples; on the other hand the discussion about the causes of vigilantism has, as Thomas Schmidt-Lux summarizes, thus far been broadly speaking between two perspectives: the assumptions about the frontier-regions by Brown and the theory of the countermovement by Kowalewski (Schmidt-Lux 2012).

Richard Maxwell Brown has been the first to address a definition of vigilantism in his historical study "Strain of Violence—Historical Studies of American Violence and Vigilantism" (Brown 1975; Schmidt-Lux 2012). Although vigilantism is not defined clearly, Brown analyzes the vigilante tradition in a classical sense as "organized, extra-legal movements" whose members take the law into their own hands (Brown 1975, pp. 95–96). Brown argues that the main reason for vigilante activities is "the absence of effective law and order in a frontier region" (Brown 1975, p. 96). As Brown does not define the frontier precisely, Ray Abrahams' (1998) ideas on it are included as well in this context. Abrahams also argues that vigilantism typically emerges in frontier zones (Abrahams 1998). For Abrahams frontiers are not always the regions around official borders (Abrahams 1998). He clarifies this when he states that "the long arm of the law does not stretch everywhere with equal force, and areas where its power is significantly diluted or resisted have a frontier quality" (Abrahams 1998, p. 24).

[3] For further literature on vigilantism see e.g. Rosenbaum and Sederberg (1976), Burrows (1977), Culberson (1990), Hitzler (1994), Pratten and Sen (2005) or Schmidt-Lux (2012, 2013a, b).

David Kowalewski developed the countermovement theory. He describes vigilantism as "the activities of private citizens, or government-employees acting off duty in their private capacity, designed to suppress deviance[4] by other citizen" (Kowalewski 2003, p. 339). Thereby, they may act individually as well as collectively and they take the law into their own hands (Kowalewski 2003). Kowalewski explains that vigilantes are building a countermovement against "growing numbers and activities of deviants" (Kowalewski 2003, p. 341). However, it is not solely the existence of those deviants, but rather how fast they grow and if they are seen as a "threatening movement" (Kowalewski 2003, p. 341). The threatening movement does not have to be criminal, but vigilantism is about pushing back movements that are seen as threatening the vigilante's way of life (Kowalewski 2003).

Digital Vigilantism: A New Phenomenon in the Digital Age

Trottier (2017) defines DV as "a process where citizens are collectively offended by other citizens' activity, and respond through coordinated retaliation on digital media, including mobile devices and social media platforms" (p. 56). Additionally, it is "a parallel form of criminal justice" that works with "weaponised visibility"—which includes but is not limited to "naming and shaming" (Trottier 2017, p. 56). This visibility is unwanted by the targeted person, and it is intense and enduring. Furthermore, persons are targeted for different reasons, and DV has different aims; e.g. from the hope for conventional justice to the targeted person losing his or her job (Trottier 2017).[5] In China, this phenomenon is called "human flesh search", and it "involves mediated search processes whereby online participants collectively find demographic and geographic information about deviant individuals, often with the shared intention to expose, shame, and punish them to reinstate legal justice or public morality" (Cheong and Gong 2010, p. 472).

[4]With deviance Kowalewski refers to "beliefs and behaviours of citizens which differ significantly from the established social norms of a community" (Kowalewski 2003, p. 339). He furthermore differentiates between criminal, cultural and political deviance (ibid.).

[5]Although this does not mean that DV is an organized process, and the different participants supposedly have different aims or a different personal motivation (also maybe without any real aim) for their participation.

Huey et al. (2012) offer a different but nevertheless important perspective on cyber-vigilantes, as they are concerned with citizens who have computer skills that allow them to get information about suspected criminals (Huey et al. 2012). Thereby, they concentrate on civilian policing on the Internet where these skilled people cooperate with the police without the "naming and shaming" and thus with the punishment that defines DV. Nhan et al. (2017) stress that cyber-vigilantes are taking more and more part in online as well as in real-world investigations (Nhan et al. 2017). This is the case for new forms of crimes, which are created through the Internet (such as hacking, distribution of destructive viruses and identity theft or cyberbullying; Nhan et al. 2017), as well as for conventional real-life crime that can be investigated online through, for instance, the identification of persons on photos, video recordings, etc.

In summary, there is digital vigilantism as defined by Trottier, which includes "human flesh search" but there is as well the phenomenon that cyber-vigilantes investigate crime and cooperate with the police. A mixture of these two phenomena is "cyber-sleuthing" (Nhan et al. 2017, p. 341) where people investigate crime online but are not cooperating with the police. Instead, they are speculating in online forums or use crowdsourcing and the knowledge of the crowd for investigations. Thereby, these phenomena can easily mix up when a cyber-sleuth or cyber-vigilante (before turning to the police) is sharing information for others to verify and look into, and hence DV can develop very easily.

For a better understanding of the concrete forms of digital vigilantism, the following sections illustrate DV examples with different characteristics. Each example is described first and then the links to conventional vigilantism are discussed. The six examples are categorized into three groups: The first two cases are incidents in which offences in real life were punished online. It is demonstrated how incidents in real life can become of interest for some Internet users. These are cases, which would not have been recognized by such a large crowd without the Internet. The subsequent two cases are online offences that are being punished online. On the one hand, it is shown how the Internet is seen as a frontier zone by some Internet users, and on the other hand it underlines that possibilities for due legal process are given and that law enforcement is of course not completely absent online. The last two cases are real-life offences that are investigated online. It is shown how the police can make use of cyber-sleuths. Thereby, the serious risks of these investigations become evident as well. The explained categorization has indeed overlaps, but it helps to highlight the different key aspects of the examples.

Real-Life Offences Attracting Online Opprobrium

The "Dog Poop Girl"

A quite famous example of digital vigilantism that exactly fits the definition of DV by Daniel Trottier is the case of a Korean woman who refused to clean up after her dog had defaecated on the floor of a subway (Dennis 2008). In reaction to the incident one of the travellers took a picture and uploaded it on a popular website (Krim 2005). Within hours the picture spread, and she was called "Dog Poop Girl"/"Dog Shit Girl" (revised romanization: Gae-Ttong-Nyeo) (Knowyourmeme.com 2005), and people looked for more information. The digilantes revealed her identity, her past and information about friends and family. People soon started to recognize her on the street, and Internet users left thousands of harsh comments on her profile on the university website. After the university took down that subpage, people started calling the university (Breitenbach 2015). The woman then reportedly left the university (Krim 2005). Korean news media reported about the "witch hunt"—as well as US bloggers (Solove 2005) and even the Washington Post (Krim 2005). They critically shed light on "the power of the Net to shame" (Dennis 2008, p. 351). The woman responded with a posted apology in which she thought about suicide (Knowyourmeme.com 2005). Today, about 13 years after the incident, it is still very easy to find information about the incident and the memes; also pictures of the woman are still available—even without anonymization.

In this case, the incident happened in real life and was predominantly punished on the Internet. Furthermore, without the Internet, it would have stayed widely unnoticed. The countermovement theory is evident, if we see urban disorder and antisocial behaviour as something people are increasingly bothered by. The urban disorder is by then the deviance, and the "dog poop girl" is the symbolic deviant for all urban disorder against which people form a countermovement. There is no real threatening movement, but in interpreting the countermovement in a broader way like this, it can explain the large participation in punishing the woman for not cleaning up.

One might slightly be reminded of the frontier theory because digilantes punished something the state did not. Certainly, it does not fit very well, as the subway is no frontier region and by stating this concept of the frontier would mean it is everywhere there is no police present.[6] However, what can be summarized is that there are people who wanted to bring justice to a woman that violated a norm with impunity. The comment concerning the incident cited in Dennis (2008) and originally posted on blogspot.com underlines this attitude (and goes even further): "Thanks to technology, we are able to build a better society in which citizens are the police, prosecutors, and judges" (Dennis 2008, p. 351).

The "Human Flesh Search Engine"

In 2008 an incident in Beijing occurred where a Chinese man violently tried to pull an 11-year-old girl into the toilet of a restaurant. She escaped and her parents confronted the man. He was unapologetic and also completely unconcerned about potential consequences his sexual harassment could have as he was "a mayor-level official appointed by the Ministry of Transport in Beijing" (Cheong and Gong 2010, p. 477). The parents of the 11-year-old girl reported the incident to Shenzhen Municipal Public Security Bureau emergency service, and somehow a local website got aware of it. It soon spread all over the Chinese Internet including videos from the restaurant's CCTV system on which the vocals were recorded as well: The physical threats against the girl's parents were audible. Only 2 days later the man was identified. Shenzhen Municipal Public Security Bureau apologized for the man's behaviour, and he was removed from his occupation (Cheong and Gong 2010). But as there were no further punishments, online discussions and demands for monetary compensation and an open apology emerged; altogether, 1.8 million comments were made in this case (Cheong and Gong 2010).

This phenomenon is an example of the effect DV can have if law enforcement is not willing to help. The fundamental of this Chinese phenomenon is the "human flesh search". This phenomenon that occurred in the last decade is still ongoing and is contested as it seems to be an effective tool (as the example above illustrates), but it can also have severe consequences and seems to be a relatively common practice in China.[7] In her plea for a legislation concerning the human flesh search, Anne

[6] This would inversely mean only in a completely controlled state a frontier would not exist.

[7] See also Tormsen (2015), Hatten (2014), Bork (2008) and Strittmatter (2010) as Chinese nationalism can also be a trigger for the start of the "human flesh search engine".

S.Y. Cheung states: "It is indeed true that the Internet has put brakes on official abuse of power, and we have made friends and foes known and unknown. But we now also know that the Internet has a grimmer side" (Cheung 2009, p. 279).

In this particular case, Chinese Internet users tried to enforce the law as the people in charge of power were not willing to—at least not at the beginning. It can be interpreted as a substitution of the state's task by citizens, and in this context evokes associations with the frontier thesis of Brown, and it is augmented by the unwillingness of the state and the aspect that citizens confronted the state with its own task. Additionally, a comment that was cited in The Guardian shows how this incident was seen in the context of the problem with corrupt officials in China, and it highlights the links to the countermovement theory: "You use public funds to eat and drink, we endure it; you show off your public cars, we endure it; you have third and fourth 'wives', we endure it; you have many villas/houses, we endure it. Now you will not even leave our own children alone" (Branigan 2008). It can be stated that people are building a countermovement against the, at least perceived, growing of deviance of authority officials.

Online Offences Attracting Online Opprobrium

The Case of Amanda Todd

The case of Amanda Todd is a very tragic one. It is also a conceptually difficult case as it mixes cyberbullying and digital vigilantism. It all started when the 12-year-old girl Amanda Todd was talked into lifting her T-Shirt by a stranger via webcam (Schütten 2012). A year later, she received messages from a man who threatened to expose the picture "if she didn't give him a show" (CBC News 2012). After she refused, her picture was sent to her family and friends. She became a victim of cyberbullying as well as bullying in school and became depressive. A couple of weeks before she finally committed suicide at the age of 15, she uploaded a YouTube video in which she described her despair (Schütten 2012).

Cyberbullying means the harassment and intimidation by users (typically youngster) of other users through digital media (Trottier 2017). According to Trottier, DV and cyberbullying are conceptually similar, as both are forms of online persecution. Yet "DV can come as a response to cyber-bullying" (Trottier 2017, p. 62). Hence, what indeed followed after the death of Amanda Todd was classical DV. Anonymous[8] targeted a man who they thought was her harasser. But Anonymous exposed the wrong person (Davison 2012). The real offender was a Dutch man who was sentenced to nearly 11 years imprisonment in 2017 for harassing altogether 34 girls and five men online (Rütten and Sendker 2017).

In this case, cyber and real-life mobbing occurred against a girl who lifted her T-Shirt online in front of a webcam. This was followed by online punishment: a classic DV campaign against an alleged offender. Anonymous tried to punish the offender online by naming and shaming, but as it has been the wrong person this case shows the very problematic aspect of DV. If Anonymous had given the information to the police instead of publishing it, the police could have started an investigation with all the rules that a due legal process demands. Instead, Anonymous jumped to conclusions and convicted a man to public humiliation. It seems like this group sees the Internet as a frontier zone. At that time the police were not able to find the offender. Thus, Anonymous tried to substitute law enforcement in trying to provide more security and safety by revealing the suspected offender's name so that he could not do any more harm.

Links to the countermovement theory are present as well, if we see the leaking in the context of online sexual harassment and paedophiles acting online. Many people consider these phenomena or people as a growing and threatening movement: Paedophile "hunting groups" that act online exist in various countries[9] and can be interpreted as a countermovement as well. In this context, Anonymous can be seen as a countermovement against this criminal deviance.

[8] Anonymous is an Internet phenomenon or a kind of community that among other things fights for the "freedom of the Internet". It does not have a structure, and everyone who acts in accordance with the aims is a member—always self-declared. Anonymous started in 2008 as a protest movement against Scientology, which they attacked via DDoS attacks. DDoS attacks are still a popular mean for Anonymous, but also hacking and apparently leaking are components of their actions (see also Patalong 2009).

[9] Mostly, those groups create fake online accounts. Then they act as a child, while chatting with potential paedophiles. When those men want to meet the "child" in real life, they meet the "paedophile hunters" whose predominant punishment is the filming of the men and the uploading of the videos online. These groups also act in other ways: while some started cooperating with the police, others are using physical violence on their targets. Examples are groups like "Letzgo Hunting" (Great Britain, see Wilson 2013), Occupy Paedophilia (a Russian, also homophobic group, see Nichols 2013) or the US series "To catch a predator" where the whole process of catching potential paedophiles was shown on a TV show that aired between 2004 and 2007 (see Hansen 2006 and Wilson 2013).

The Case of Zane Alchin

In the case of Zane Alchin, DV ended with a conviction. A friend of Zane Alchin posted on Facebook the screenshot of the Tinder profile of Olivia Melville who used a rap lyric saying “Type of girl that will suck you dry and then eat some lunch with you” in her description (Smith 2016). The screenshot was shared thousands of times, vitriolic, misogynist and threatening comments followed; Zane Alchin commented in this way as well (Vernon 2016).

As the profile was not meant for a large social media audience, the incident has indeed a naming and shaming aspect and can be classified as DV. Here, some men were obviously of the opinion that women should not make sexual references and shamed her for doing so. Paloma Brierley Newton defended Olivia Melville and became herself a target of Zane Alchin’s comments (Vernon 2016). She then took screenshots of the comments and handed them to the police. Zane Alchin was charged and given a 12-month good behaviour bond by the court. This shows that law enforcement is indeed possible. However, the process was still accompanied by online harassment now targeting Zane Alchin himself, and the defence attorney stated that this online harassment was a punishment far beyond every conviction a court would have imposed (Kembrey 2016). The Zane Alchin case does fit the DV definition as he and other men named and shamed Olivia Melville.

In other cases, women name and shame publicly without turning to the police but for example to the employer of an online harasser causing the loss of his job. Emma Jane writes: “Feminist digilantism returns power to female targets, raises awareness, and makes internet attackers accountable for their actions when institutions have failed” (Jane 2016a). But she also states that it has risks (innocent people getting hurt, becoming scapegoats, etc.) and that it is “a symptom of, not a solution to, online misogyny” (Jane 2016a).

“Feminist digilantism”—as Jane calls it—is by some women seen as necessary when there is no sufficient law enforcement procedure; here, the Internet is seen as a frontier zone in which women try to provide safety and security for each other. Concerning this particular case, by turning to the police one woman showed how law enforcement could be brought to the frontier zone.

In the Zane Alchin case as well as in the general “feminist digilantism” context we can observe a countermovement: Women form a countermovement to deviant behaviour of misogynistic men in particular and sexism in general. Especially with the ongoing #metoo-debate[10] about sexual harassment, this countermovement seems to grow, and digilantism as a response to harassment—online and in real life—could likely be a practice that goes on.

Real-Life Crime Being Investigated Online

The Vancouver Riot (2011)

A well-researched incident that directly involved police work is the 2011 Vancouver riot in which up to 100,000 people were involved. In consequence of the violent outbreaks cars were burnt and other violent acts were committed (Trottier 2012). In reaction to the violence, people started recording the events via mobile devices. They uploaded the content on a Facebook group called "Vancouver Riot Pics: Post Your Photos" which was built to share the material. The police used the photos and videos to identify rioters and with the help of users who "tagged" rioters, the identification process was even easier (Schneider and Trottier 2012). This procedure represented "a highly successful use of crowdsourcing by the police" (Nhan et al. 2017, p. 358). Apart from that it also had the effect that the police could not "define the situation by providing narratives of the event" (Schneider and Trottier 2012, p. 59), as it was the case in 1994 where a similar riot occurred. Back then, the police confiscated evidence from different media that filmed the riot also documenting the use of force by the police. The parts, which depicted this use of force by the police, were removed before the videos were shown in public so that rioters could be identified (Schneider and Trottier 2012). However, in the 2011 riot where social media were crucial to the identification of rioters, "vitriolic reactions" occurred as well and those evoked association with a mob mentality (Schneider and Trottier 2012, p. 60).

Contrary to some of the other cases, this event would have been noticed without the Internet: A similar previous riot were widely covered by the news, but the search for the rioters in 2011 was much easier as they became more visible. A sergeant concerned with the riot stated that the police

[10] The #metoo debate occurred in October 2017 when women were invited to post #metoo on social media if they had ever been a victim of sexual harassment. In this context, quite a lot of prominent men in showbusiness have been accused of sexual harassment and lots of them admitted it. Right now, it is debated how society should cope with that.

"have names of the suspects before we even know exactly what they did and where they did it" (cited in Trottier 2012, p. 416). These cases partly fit into the definition of DV by Daniel Trottier because DV was one part, but the events are evoking more associations with cyber-vigilantes described by Huey et al. (2012) or cyber-sleuths. Real-life offences happened and offenders could be found through the Internet with the help of Internet users (Huey et al. 2012).

Furthermore, there was a partly frontier region as the police were not but should have been prepared for the violence (Lee 2011). A threatening movement existed as well against which Internet users built a countermovement. Furthermore, the support of the police by the digilantes needs to be highlighted. There were harsh comments against some of the alleged rioters, whose impact may be by no means without complications, but still people helped the police by tagging the rioters. It seems that by this support they built a countermovement trying to balance the temporarily partial frontier zone.

The Boston Marathon Bombings

The terror attack at the Boston Marathon on 15th April 2013 caused three deaths and injured 264 people (CNN Library 2017). The bombings were followed by "one of the largest, most sweeping investigations and manhunts in US history" (Nhan et al. 2017, p. 341). A 4-day long search for the terrorists ended with one dead and the other one captured. The latter has been sentenced to death in 2015 (CNN Library 2017). In the process of finding the terrorists, a large online community participated in the investigations. These "cyber sleuths were pooling information and resources in order to assist the police in their criminal investigation of the bombing" (Nhan et al. 2017, p. 341).

Nhan et al. (2017) analyzed discussion threads on Reddit in the aftermath of the bombings and concluded that there is high potential in crowdsourcing information as the people commenting there provide quite a lot of skills and expertise (Nhan et al. 2017).[11] In parts the FBI turned to the crowd, initially requesting to share images and videos with them (Ngak 2013). Later they also requested the identification of the two suspects when they released photos of the later killed/arrested terrorists (Tapia

[11] This is underlined by the example of thread members who gave detailed information about the type of the bomb (Nhan et al. 2017, pp. 348f).

et al. 2014). There were a large number of Reddit users who sent information to the police (Nhan et al. 2017). Yet the FBI also had to release information earlier than planned because of the crowd's pressure (Tapia et al. 2014).

Furthermore, the investigations of the Reddit users became very problematic as they identified wrong suspects. The 17-year-old Salah Barhoun was tagged in a picture online, which went all over social media and was later on publicized by "The New York Post" (Rahmanzadeh et al. 2013). Another suspect accused by Reddit users was Sunil Tripathi, and in consequence his mother received threatening messages. Later, it turned out that he committed suicide before the bombings happened (Kang 2013; Nhan et al. 2017). There were further threads with much misleading information about "suspects" and "suspicious" behaviour[12] (Nhan et al. 2017).

Colin Hill summarizes the Boston bombing digilantism: "Cyber-vigilantism, ironically, exacerbated the chaos it was meant to resolve, with misidentified suspects and inaccurate information blurring the line between truth and mere speculation" (Hill 2017). Although crowdsourcing has potential, as using the knowledge of the crowd can lead to better investigation results, it implies risks. Information are immediately public and not everyone follows the presumption of innocence at all times—indeed as was seen, there are just too many people condemning suspects who ultimately did not participate in any sort of crime. The "cyber-sleuthing" in the aftermath of the Boston Marathon Bombing was widely discussed in media, and there are initial approaches to this phenomenon in different fields of research (Marx 2013; Nhan et al. 2017).

Nhan et al. (2017) summarize that "online communities of 'digilantes' are here to stay" (Nhan et al. 2017, p. 358). With this knowledge they should be used effectively, because „a lack of guidance and active feedback from law enforcement increases the likelihood that such efforts are redundant or subject to error" (Nhan et al. 2017, p. 353). DV campaigns as described by Trottier were part of the user-led investigations in the aftermath of the Boston Bombings—as the threatening messages to the suspect's mother displays. Are there links to the vigilantism theories though? Boston is no frontier region—especially proven by the police catching the terrorist not Reddit users. It seems that in this case cyber-sleuths were driven by the willingness to help the police and get involved in crime investigations, which was not an option before the possibilities of social

[12] An overview over the speculations is given by Chen (2013).

media. A link to the countermovement theory is given insofar as in the case of terrorism, which can be framed as a growing threatening movement, people are building a countermovement against the perceived threat. This could partly explain the large willingness to help and not to let "just" the police do their work.

DV as an Extension of Vigilantism

Examples on DV are manifold and they are rarely without problematic aspects. Further examples can be easily added: from anti-paedophilia campaigns to Facebook groups in which people upload photos of badly parked vehicles to controversial tweets and comments that go viral and trigger DV.

The compilation in Table 7.1 gives an impression about different aspects of DV and the six concrete examples.

The compilation above has indeed overlaps and is not as clear as the table suggests. However, it underlines the main differences between the cases, and it can be stressed that these different types of DV need different reactions by the authorities.

When there is real-life punishment by vigilantes, it might be easier for the police to react to the actual incident as well as to the behaviour of the vigilantes (if the incident/vigilante behaviour is illegal). Still, the online

Table 7.1 Overview of different aspects of DV

Incident	*Occurrence*	*Punishment*	*Consequences*	*Classification*
Dog poop girl	Real-life	Online	Humiliation	Real-life offences being punished online
Human flesh search against "mayor level official"	Real-life	Online	Humiliation/job loss	
Amanda Todd	Online	Online	Humiliation/ suicide /legal process	Online offences being punished online
Zane Alchin	Online	Online	Humiliation/ legal process	
Vancouver riot	Real-life	Online	Humiliation/ legal process	Real-life crime being investigated online
Boston marathon bombings	Real-life	Not successful in finding real terrorists	Humiliation for wrong suspects	

behaviour/punishment may be as bad as real-life vigilantism; accordingly, more work on how to react to the phenomenon must be done. In those cases where the incident occurred online, the Internet appears as a frontier zone in which people are trying to substitute an absent law enforcement—even though in the two examples legal trials followed (although years later in the Amanda Todd case). Thereby it can be shown that although the Internet is partly a frontier zone, the existing constitutional states do have the means to punish online offences or online crimes via legal process—which should also show participants of DV that cooperation with the police is certainly the better way than digilantism.[13]

Despite that, not all DV is a reaction to real offences; there is also DV occurring, e.g. against a tweet that is interpreted as a norm violation. Here, law enforcement is not in charge. Thus, law enforcement will not be able to stop DV, but at least in cases where activities are relevant under criminal law it should be active. It is especially important not to ignore digital vigilantism just because it may not always be physical harm done at first sight. It seems that DV is often seen as an effective tool, at least when law enforcement is not able or willing to pursue, e.g. in case of online harassment.

For the last grouping "real-life crime being investigated online", Nhan et al. (2017) should be cited. They argue that digilantes are here to stay. Therefore, it is important to find ways to manage or control them—on the one hand because the police might even profit from efforts as in the case of the Vancouver riot, on the other hand because preventing that false suspicions going viral seems to be central for those kinds of DV.[14]

The discussion of the six examples above underlines further aspects. The very heterogeneous DV cases have in common that the punishment has mainly been online. This punishment was executed via humiliation. Although, the humiliating process often had different characteristics, the naming and shaming aspect and the "weaponization of visibility" (Trottier 2017, p. 56) are essential.

These insights are not surprising as they are inherent to DV. It is to highlight that no matter whether the incident happened in real life or online, the consequences are similar regarding humiliation, as the online punishment can have intense real-life consequences for those targeted. Especially—but

[13] At least if it is possible; in the case of the Chinese official cooperation with the police might not have been a promising strategy.

[14] Furthermore, if in any future case cyber-sleuths are successful in finding a real criminal offender, the possibility of vigilante justice might be higher.

not only—in cases of false suspicion, the missing accountability is central and the question of who is responsible if something goes wrong occurs: everyone who participated or rather the ones commenting "really badly". However, these participants or commentators could be found out. This leads to a lot of, mainly legally relevant follow-up questions.[15]

Despite those questions, the large-scale participation is critical with regard to two points. Firstly, it is (almost) impossible for law enforcement to prosecute everyone who violated, for instance, privacy rights. Here again, legal issues are central such as the different legislations in different countries. Secondly, the response time online is quite short, and the more commentaries are made the more likely it is that the whole process is getting out of hand[16]—in other words, often some kind of self-reinforcing tendencies can be observed.[17]

Often, the proportionality of the punishment that is carried out by many people is questionable, even if the target is the correct person and not wrongly suspected. Compared to conventional vigilantism, at this point digital vigilantism differs with regard to the number of people becoming aware of the (alleged) misbehaviour. Still, these two phenomena should not be seen as outright different practices, which leads to the initial question of this article: After the overview over the different aspects and the fact that from the disparate groups of DV examples, different consequences occur, the questions raised in the beginning—whether there is a shift from vigilantism to digilantism or a continuing existence of conventional vigilantism through new means—should be addressed.

Although the theories of Brown and Kowalewski are broadened for some cases to explain the occurring DV behaviours, all cases have links to one or both of the theories. Additionally, as vigilantes, digital vigilantes are often trying to provide safety (although safety interpreted very broadly in some cases) by fighting disliked or criminal behaviour and some are taking the law into their own hands as well. It can furthermore be argued that there are

[15] The legislation that is needed for due legal process concerning commentaries and so on comes with problems that affect, for instance, freedom of speech, and is not that easy itself. This can be observed in Germany where a new legislation is strongly criticized (see for example Reinbold 2017). Furthermore, DV participants often come from different countries, which have different legislation.

[16] See also Halberschmidt (2014) who argues that mass protests have been there before the Internet, but online protests (or in this case DV) are much larger and faster.

[17] Here, the links to other Internet phenomena like shitstorms (which I would argue can indeed be a part of DV) are clear. For general principles on social media and how they work see for example Schmidt (2013).

digital vigilantes that aim at the preservation of their norms and values. Also, as vigilantism, DV works in a moral and legal grey area (Arfsten 2012).

When Anne S.Y. Cheung writes about the human flesh search engine, the links become even more evident: "The mob, infuriated at various objectionable practices one has engaged in, *may take the law into their hands and bring 'justice' to its end*" (Cheung 2009, p. 279, emphasis by the author). The same is true for the statement of Emma Jane when she writes about law enforcement concerning online abuse, using exactly the words of Ray Abrahams: "After many years of lacklustre response, *the long arm of the Australian law is finally extending to the cybersphere*" (Jane 2016b, emphasis by the author).

It has been shown that theories on conventional vigilantism are still adequate to explain DV at least partly. Therefore, existing theories of vigilantism should not be forgotten as they can give different perspectives on DV. Digital vigilantism with all its problematic aspects seems indeed to augment conventional vigilantism. Those two phenomena should consequently not be seen as two outright disparate practices, although there are differences. Like other social phenomena appearing in real life as well as online, digital vigilantism in some cases mixes up real-life activities with social media, blurring the online–offline distinction.

Coming back to the question whether DV is simply vigilantism through new means, the answer is indeed "yes" to a great extent. But it must be stressed that these "new means" are influencing digital vigilantism: Social media have special characteristics that influence the new form of vigilantism causing it to become this new practice of DV we are observing and should have a closer look at. On the one hand, through social media, misbehaviour can become more visible and be punished more easily. On the other hand, social media or the Internet itself can be the source for activities that are then punished via social media, especially because social media as a partly frontier zone seem to trigger some norm violations.

It is helpful to know about digital vigilantism and to know about its severe consequences, so that reactions can be developed for the different kinds of DV. For this, it might be helpful to take a global perspective because DV often does not remain within one country. Overall, it can be stated that the framing of DV in the theories of vigilantism is—despite difficulties[18]—possible and useful. Additionally, scientific efforts to get, for instance, a typology that leads to a deeper understanding of DV are on

[18] Not only because vigilantism theories are diverse, but also because DV comes in different forms.

their way, and this is indeed necessary: With the developments in social media and a world that is increasingly connected via those platforms, different forms of digital vigilantism are more likely to grow than to decline. Finally, there is no shift from vigilantism (which still exists in its conventional form) to digilantism but rather an extension of vigilantism.

References

Abrahams, R. (1998). *Vigilant citizens: Vigilantism and the state*. Cambridge: Polity Press.

Arfsten, K.-S. (2012). "Every man is entitled to defend his castle…" - Vigilantismus während der London Riots. *Kriminologisches Journal, 44*(2), 101–117.

Bork, H. (2008, April 19). Vorolympische Hexenjagd auf selbstkritische Chinesen. *Tagesanzeiger*. Retrieved November 30, 2017, from https://archive.is/20121015171253/http:/sc.tagesanzeiger.ch/dyn/news/zuerich/863350.html

Branigan, T. (2008, November 5). Chinese official sacked over attack on girl, 11. *The Guardian*. Retrieved November 30, 2017, from https://www.theguardian.com/world/2008/nov/05/china-lin-xiaxiang-communist-party.

Breitenbach, P. (2015). Netiquette: Auf der Suche nach einer digitalen Etiquette. In C. Stiegler, P. Breitenbach, & T. Zorbach (Eds.), *New media culture: Mediale phänomene der netzkultur. Digitale gesellschaft* (Vol. 5, 1st ed., pp. 143–162). Bielefeld: Transcript.

Brown, R. M. (1975). *Strain of violence: Historical studies of American violence and vigilantism*. New York: Oxford University Press.

Burrows, W. E. (1977). *Vigilante!* New York: Harcourt Brace Jovanovich.

CBC News. (2012, October 16). Amanda Todd's alleged tormentor named by hacker group. *CBC News Online*. Retrieved October 23, 2017, from http://www.cbc.ca/news/canada/british-columbia/amanda-todd-s-alleged-tormentor-named-by-hacker-group-1.1134233.

Chavez, L. R. (2005). Spectacle in the desert: The minuteman project on the US-Mexico border. In D. Pratten & A. Sen (Eds.), *Global vigilantes* (pp. 25–46). New York: Columbia University Press.

Chen, A. (2013, April 17). *Your guide to the Boston Marathon bombing amateur internet crowd-sleuthing*. gawker.com. Retrieved from http://gawker.com/5994892/your-guide-to-the-boston-marathon-bombing-amateur-internet-crowd-sleuthing

Cheong, P. H., & Gong, J. (2010). Cyber vigilantism, transmedia collective intelligence, and civic participation. *Chinese Journal of Communication, 3*(4), 471–487.

Cheung, S. A. (2009). China internet going wild: Cyber-hunting versus privacy protection. *Computer Law & Security Review, 25*(3), 275–279.

CNN Library. (2017, March 29). Boston Marathon terror attack fast facts. *CNN*. Retrieved October 25, 2017, from http://edition.cnn.com/2013/06/03/us/boston-marathon-terror-attack-fast-facts/index.html.

Culberson, W. C. (1990). *Vigilantism: Political history of private power in America: Contributions in criminology & penology*. New York: Greenwood Press.

Davison, J. (2012, October 22). Online vigilantes: Is 'doxing' a neighbourhood watch or dangerous witch hunt? *CBC News Online*. Retrieved October 27, 2017, from http://www.cbc.ca/news/technology/online-vigilantes-is-doxing-a-neighbourhood-watch-or-dangerous-witch-hunt-1.1132015.

Dennis, K. (2008). Keeping a close watch—The rise of self-surveillance and the threat of digital exposure. *The Sociological Review, 56*(3), 347–357.

Halberschmidt, T. (2014, October 3). Entstehung eines Shitstorms: Warum die Medien eine Mitschuld haben. *Handelsblatt*. Retrieved January 12, 2018, from http://www.handelsblatt.com/panorama/aus-aller-welt/entstehung-eines-shitstorms-warum-die-medien-eine-mitschuld-haben/10778756.html.

Hansen, C. (2006, October 26). They're still showing up. *NBC News*. Retrieved December 29, 2017, from http://www.nbcnews.com/id/14824427/ns/dateline_nbc/t/theyre-still-showing/#.WkYKL9_ibZt.

Hatten, C. (2014, January 28). China's internet vigilantes and the 'human flesh search engine'. *BBC News*. Retrieved November 30, 2017, from http://www.bbc.com/news/magazine-25913472

Hill, C. (2017). How digital communities cope: Cyber-vigilantism following the Boston Marathon bombings. *The Yale Review of International Studies, 7*(2), 5–11.

Hitzler, R. (1994). Die neuen Vigilanten: über Formen der Bewältigung alltäglicher Verunsicherung. *Bürgerrechte & Polizei, 48*(2), 67–71.

Huey, L., Nhan, J., & Broll, R. (2012). 'Uppity civilians' and 'cyber-vigilantes': The role of the general public in policing cyber-crime. *Criminology & Criminal Justice, 13*(1), 81–97.

Jane, E. (2016a, April 11). DIY internet justice is a symptom, not a solution to online misogyny. *The Sydney Morning Herald*. Retrieved November 30, 2017, from http://www.smh.com.au/lifestyle/news-and-views/opinion/diy-internet-justice-is-a-symptom-not-a-solution-to-online-misogyny-20160410-go2z6z.html.

Jane, E. (2016b, July 14). Stopping online abuse isn't censorship it's the least we can do. *The Sydney Morning Herald*. Retrieved November 30, 2017, from http://www.smh.com.au/comment/stopping-online-abuse-isnt-censorship-its-the-least-we-can-do-20160713-gq4oj5.html-can-do-20160713-gq4oj5.

Kang, J. C. (2013, July 25). Should Reddit be blamed for the spreading of a smear? *New York Times Magazine*. Retrieved October 26, 2017, from http://www.nytimes.com/2013/07/28/magazine/should-reddit-be-blamed-for-the-spreading-of-a-smear.html.

Kembrey, M. (2016, July 30). Sydney labourer Zane Alchin sentenced for harassing women on Facebook. *The Sydney Morning Herald*. Retrieved November 30, 2017, from http://www.smh.com.au/nsw/sydney-man-zane-alchin-sentenced-for-harassing-women-on-facebook-20160728-gqg8l8.html.

Knowyourmeme.com. (2005). *Dog poo girl*. Retrieved October 23, 2017, from http://knowyourmeme.com/memes/dog-poo-girl.

Kowalewski, D. (2003). Vigilantism. In W. Heitmeyer & J. Hagan (Eds.), *International handbook of violence research* (pp. 339–349). Dordrecht: Kluwer Academic Publishers.

Krim, J. (2005, July 7). Subway fracas escalates into test of the internet's power to shame. *Washington Post*. Retrieved October 23, 2017, from http://www.washingtonpost.com/wp-dyn/content/article/2005/07/06/AR2005070601953.html?referrer=emailarticle.

Lee, J. (2011, September 1). *Police, city poorly prepared for vancouver stanley cup riot, says report*. Vancouver Sun. Retrieved October 26, 2017, from http://www.vancouversun.com/news/Police+city+poorly+prepared+Stanley+riot+Report/5341286/story.html.

Marx, G. T. (2013). The public as a partner? Technology can make us auxiliaries as well as vigilantes. *IEEE Security & Privacy, 11*(5), 56–61.

Ngak, C. (2013, April 17). Crowdsourcing or witch hunt? Reddit, 4chan users try to ID Boston bomb suspects. *cnet*. Retrieved October 26, 2017, from https://www.cnet.com/news/crowdsourcing-or-witch-hunt-reddit-4chan-users-try-to-id-boston-bomb-suspects/.

Nhan, J., Huey, L., & Broll, R. (2017). Digilantism: An analysis of crowdsourcing and the Boston Marathon bombings. *British Journal of Criminology, 57*(2), 341–361.

Nichols, J. (2013, September 5). 'Occupy Paedophilia,' Russian anti-gay group, 'on safari' hunting gays (VIDEO). *The Huffington Post*. Retrieved December 29, 2017, from https://www.huffingtonpost.com/2013/09/05/occupy-paedophilia-hunting-gays-russia_n_3875223.html.

Patalong, F. (2009, September 14). Anonymer Angriff aus dem web. *Spiegel Online*. http://www.spiegel.de/netzwelt/netzpolitik/protestkultur-anonymer-angriff-aus-dem-web-a-648653.html.

Pratten, D., & Sen, A. (Eds.). (2005). *Global vigilantes*. New York: Columbia University Press.

Rahmanzadeh, S., Leong, J., Riley, S., Leuci, S., & Schwartz, R. (2013, April 18). Teen: I am not the Boston Marathon bomber. *abc news*. Retrieved October 26, 2017, from http://abcnews.go.com/Blotter/teen-boston-marathon-bomber/story?id=18990057#.UXBs_CtASOv.

Reinbold, F. (2017, June 30). Das Facebook-Gesetz ist erst der Anfang. *Spiegel Online*. http://www.spiegel.de/netzwelt/netzpolitik/heiko-maas-und-facebook-das-netzdg-ist-erst-der-anfang-die-analyse-a-1155222.html.

Rosenbaum, H. J., & Sederberg, P. C. (Eds.). (1976). *Vigilante politics*. Philadelphia: University of Pennsylvania Press.

Rütten, F., & Sendker, J. (2017, March 22). Wie ein Cyber-Stalker ein junges Mädchen in den Tod trieb. *Stern Online*. Retrieved November 24, 2017, from https://www.stern.de/panorama/stern-crime/fall-amanda-todd%2D%2Dwie-ein-cyber-stalker-ein-junges-maedchen-in-den-tod-trieb-7379148.html.

Schmidt, J.-H. (2013). *Social media*. Wiesbaden: VS Verlag für Sozialwissenschaften.

Schmidt-Lux, T. (2012). Vigilantismus. Ein Phänomen der Grenze? *Kriminologisches Journal, 44*(2), 118–132.

Schmidt-Lux, T. (2013a). Jenseits von Batman. Schlüsselfiguren des Vigilantismus. *Forschungsjournal Soziale Bewegungen, 26*(4), 64–71.

Schmidt-Lux, T. (2013b). Vigilantismus als politische Gewalt. Eine Typologie. *BEHEMOTH A Journal on Civilisation, 6*(1), 98–117.

Schneider, C. J., & Trottier, D. (2012). The 2011 Vancouver riot and the role of Facebook in crowd-sourced policing: Research note. *The British Columbian Quarterly, 175*(Autumn), 57–72.

Schütten, S. (2012, October 25). Cybermobbing lässt sich nicht allein im Netz bekämpfen. *Zeit Online*. Retrieved October 23, 2017, from http://www.zeit.de/digital/internet/2012-10/amanda-todd-anonymous.

Smith, J. (2016, June 20). 'Slut shamer', 25, could face three years' jail after pleading guilty to abusing a woman on Facebook after she quoted drake lyrics on her Tinder profile. *Mail Online*. Retrieved November 30, 2017, from http://www.dailymail.co.uk/news/article-3649867/Zane-Alchin-faces-jail-abusive-Facebook-comments-raping-feminists-woman-included-explicit-Drake-lyric-Tinder-profile.html.

Solove, D. (2005). *Of privacy and poop: Norm enforcement via the blogosphere*. Retrieved October 23, 2017, from https://balkin.blogspot.de/2005/06/of-privacy-and-poop-norm-enforcement.html.

Strittmatter, K. (2010, May 11). Die Menschenfleischsuche. *Süddeutsche Zeitung*. Retrieved November 30, 2017, from http://www.sueddeutsche.de/politik/trainieren-fuer-olympia-die-menschenfleischsuche-1.204850.

Tapia, A. H., LaLone, N., & Kim, H. W. (2014). Run amok: Group crowd participation in identifying the bomb and bomber from the Boston marathon bombing. In ISCRAM 2014 Conference Proceedings - 11th International Conference on Information Systems for Crisis Response and Management (pp. 265–274). The Pennsylvania State University.

Tormsen, D. (2015). *10 stories of China's human flesh search engines*. Listverse.com. Retrieved November 30, 2017, from https://listverse.com/2015/04/22/10-stories-of-chinas-human-flesh-search-engines/.

Trottier, D. (2012). Policing social media. *Canadian Review of Sociology, 49*(4), 411–426.

Trottier, D. (2017). Digital vigilantism as weaponisation of visibility. *Philosophy & Technology, 30*(1), 55–72.

Vernon, J. (2016, June 20). Zane Alchin pleads guilty to making Facebook threats about Olivia Melville Tinder profile. *ABC News*. Retrieved November 30, 2017, from http://www.abc.net.au/news/2016-06-20/zane-alchin-pleads-guilty-to-facebook-abuse-of-olivia-melville/7525664.

Wilson, D. (2013, April 22). Letzgo hunting and the dangers of paedophile vigilantism. *The Guardian. Retrieved from* https://www.theguardian.com/commentisfree/2013/apr/22/paedophile-vigilantism.

CHAPTER 8

Building a Bedrock of Trust Between Citizens, Law Enforcement and Other Stakeholders: A Scalable Architecture for Community Policing

Laurence Marzell and Ben Brewster

Introduction

Community policing (CP), as a concept, is a policing approach that seeks to establish working partnerships between the police and the communities they serve (Peak and Glensor 1996). It is an extension of the social contract that exists between police and citizens, and an approach that seeks to establish collective efficacy within the community (Brewster et al. 2018; Rukus et al. 2018). CP strategies are designed to empower residents, allowing them to participate in discussions around policing, and to work more closely with frontline officers in their neighbourhoods (Community

L. Marzell (✉)
Serco Europe, Hampshire, UK
e-mail: Laurence.Marzell@serco.com

B. Brewster
CENTRIC, Sheffield Hallam University, Sheffield, UK
e-mail: B.Brewster@shu.ac.uk

B. Akhgar et al. (eds.), *Social Media Strategy in Policing*,
Security Informatics and Law Enforcement,
https://doi.org/10.1007/978-3-030-22002-0_8

Oriented Policing Services 2014). Popular initiatives such as neighbourhood watch, citizens advice panels and police operated youth programs are all based around the core values of CP (Maguire et al. 1997). While this partnership approach between police, citizens and other statutory authorities has become a fairly widespread and accepted approach, actually developing and maintaining these collaborative relationships is often a complex and context sensitive issue. Law enforcement culture can make empowering citizens and creating collective efficacy difficult, and CP approaches have long been maligned for botched implementations that provide little in the way of citizen empowerment (Reisig and Parks 2004).

In today's complex and interconnected world, the diverse communities that exist within it are complex ecosystems; no longer defined by geography alone. Instead, a multitude of factors, interests and connections, both virtual and physical, bind us all together. These connections define the actions of the aforementioned approaches that seek to establish collaboration between citizens, the police and other statutory organisations. Indeed, close, mutual, and trusted collaboration between these agencies and organisations, and the citizens and communities they serve, is an essential component in ensuring the fulfilment of their duty to protect and promote the wellbeing of these citizens and communities.

While CP manifests itself very differently to that of other forms of policing and law enforcement, it is information that connects them. Information, its use and governance, enables valuable and meaningful CP. It is able to build and sustain trust across and within citizens, communities and law enforcement. Information flows, however, are not fluid like water. Organisational boundaries and the many different rules and requirements that apply mean that information is often restricted or prevented from being used by those most relevant, or in a way that can do most good.

A contributory cause is that organisations often view the world in which they operate very differently from one another. These differing views are driven by many factors, such as: risk, history, culture, capability, economics and leadership. These factors fuel how organisations conduct their business: their governance and policies, training, budgets, processes and systems.

These differences, especially in governance and policy, where resulting information and decisions need to flow across organisational, operational or jurisdictional boundaries, both internally and externally, are significant areas of risk. Failures can, and often do, occur (cp. Chap. 3). This is especially true in our interconnected world where dependencies and interdependencies can, in the wake of a major incident, lead to consequences and cascading effects. Collaboration and communication with citizens and communities can be complex, but it is essential in CP where a user-centred approach is critical to achieving desired CP outcomes. These outcomes rely on a unified, coherent and cohesive view of roles, responsibilities and capabilities; relative to the use of information.

In this chapter, a community policing architecture framework (CPAF) is defined. The CPAF framework provides a common and shared understanding of the governance, policy and flow of information, across and between the multiple stakeholder organisations involved. It includes not only citizens and law enforcement but all stakeholder actors. This creates a mechanism to better understand and share information at a community level, providing an essential building block to create and sustain trust and, ultimately, the means by which shared CP outcomes can be achieved.

Community Policing Outcomes

CP across the UK, Europe and further afield, is the bedrock for keeping society and its communities and citizens safe, secure and resilient. Perhaps the most prominent contemporary context for the potential value of CP in recent times is that of Terrorism. Attacks, and their aftermath, have had a significant impact upon communities around the world in recent years. These events have placed a particular spotlight the values that underpin CP, exemplified by the two examples provided below.

Belgium

The formal introduction of "community-oriented policing" (COP) as the official philosophy of policing in Belgium was introduced as part of drastic reform in 1998. Under this reform, COP provided a policing strategy focused on building ties and working closely with members of communities. COP sought to create partnerships between law-enforcement agencies and other organisations, such as government agencies, community members, non-profit service providers, private businesses and the media. These partnerships aimed to provide a powerful channel through which the police could communicate with the public. Building on COP, CP recognises that the police cannot solve every public safety issue alone. Instead, interactive partnerships are created in which the public assists police in developing problem-solving solutions. In Belgium, the approach was advocated by all political parties, majority and opposition. Finally, the adoption of COP was consolidated in a circular letter distributed by the minister of internal affairs in 2003 (Renard et al. 2016).

The USA

In the USA, a community outreach officer—Jim Buck of the Los Angeles Police Department (LAPD)—uses community policing strategies to fight terrorism. The strategy was called for by the US Department of Homeland Security in order to prevent the spread of violent extremism. The LAPD's CP work goes far beyond that of Officer Buck, however. Community policing forms part of a broader strategy of partnership work between law between law enforcement and the communities they police. These techniques emphasise proactive, joint problem-solving to build trust and cooperation and address the conditions that diminish public safety. Nearly one half of policing agencies in the USA are using the community policing practices of outreach and engagement to communities being targeted for recruitment to terrorism, according to a recent national study led by Duke University's David Schanzer (Weine 2017).

But terrorism, like the other issues and challenges facing society at the grassroots, can be proactively tackled before it escalates, through effective CP practices. The framework proposed in this chapter towards building effective practices is, in part, informed by the research conducted as part of the Unity project. Unity was an EU Horizon 2020-funded project aiming to enhance the relationships and communication channels that exist between police and the communities they serve (Vickers et al. 2018). To this end, the project aimed to identify CP best practices, and use those practices to develop technological solutions to improve the communication between police and communities (cp. Chaps. 4 and 6), while also using them as a foundation for the development of police training and awareness around CP. Through further background research and a number of interviews with existing CP stakeholders, this evolved into three important CP concepts:

- Working together, cooperation and collaboration between police and external groups
- Building relationships of trust, confidence and understanding between police and external groups
- Addressing local needs and issues, both proactively and reactively (van der Giessen et al. 2016, p. 132).

Primary research and requirements elicitation exercises were undertaken in eight participating countries: Belgium, Bulgaria, Croatia, Estonia, Finland, Germany, Macedonia and the UK. Interviews were conducted with police, young minority people, intermediaries (who work with young minorities and the police) advocates, and legal and academic experts (Bayerl et al. 2015). From this research, the project developed its work based on the identification of six pillars of CP (see Akhgar et al. (2017) for a summary of the six pillars):

- Trust and confidence building
- Accountability
- Information sharing
- Crime prevention
- Addressing local needs
- Working in a collaborative manner

To achieve these shared outcomes across a range of multiple and diverse stakeholders and organisations, including citizens and community stakeholder actors, is challenging. Each views the world very differently. These views are derived from their experiences, history and culture, and factors such as risk, capability, economics, legislation and leadership. From these views and factors flow how organisations conduct their business—their governance and policies, their training, their budgets, as well as the processes, systems and technology that underpin their operations and how they carry out their CP approaches and relationships on the ground (Marzell et al. 2017a, b).

An organisation's view of the world, relative to the other organisations with which it must collaborate, is neither right nor wrong and neither good nor bad; it is just different. And all these views have value. But these differences, especially at the interchange of organisational boundaries, whether internal or external, are where gaps exist and risks arise. Information and its flow between and across these different organisations and stakeholder groups, so critical in creating a shared understanding of the issues and challenges faced, needs a framework to bridge these gaps and to build the Trust so essential to achieve the desired CP outcomes on the ground.

Former British prime minister and famed novelist Benjamin Disraeli (1880) famously declared that the most successful people in life are generally those who have the best information. Still holding true today, this statement applies equally to communities seeking to understand and meet shared issues and challenges. Information, its flow and use across and between the many diverse stakeholder actors and organisations, provides the currency and lubrication needed to improve CP. And it is this shared information and its use that shows what works well and what does not in current CP practices at a community level; and how that information can be applied to innovate and improve new CP models and approaches to meet today's dynamic and complex challenges.

In this chapter, we establish a community policing architecture framework (CPAF). The framework targets the use of information as an enabler of CP, underpinning a shared view of risks, threats, vulnerabilities and hazards across community ecosystems. This scenario-driven view supports a governance framework that enables the many organisations involved to work more effectively together. Collaborating with the citizens and communities they serve, through a better understanding, sharing and management of information to address the challenges within the complex community ecosystem.

The Community Policing Architecture Framework (CPAF)

Enterprise architecture (EA) is defined as "The fundamental organization of a system, embodied in its components, their relationships to each other and the environment, and the principles governing its design and evolution" (ISO/IEC/IEEE42010:2011 2011). The CPAF is built based upon the core principles of EA. The CPAF simplifies and manages the inherent complexity dynamic of multi-stakeholder environments. As a "single source of truth" it drives agile and iterative testing and governs the rules and principles of CP across multiple different partner organisations and stakeholder actors. It is an architecture to capture, unify and clarify governance and policy through to operational implementation, with a shared approach to improving CP through collective decision-making. This creates an environment in which trust can flourish and improved CP outcomes are achieved.

For each of the eight locations used as part of the research that contributed to this chapter we captured, held and structured information within the CPAF. This includes what information is used, how it is used and the constraints of its use across and between different organisations involved in CP at the community level. This creates a Current Operating Model (COM) of CP. From this, scenarios were used to build a set of detailed process flows within the architecture framework to visualise the activities that take place. Scenarios were contextual and designed to replicate common or "wicked" CP challenges that were prevalent in each of the pilot locations. The CPAF enables understanding at increasing levels of granularity, how joint decision making in the context of CP occurs: the processes, procedures and systems that support it, the governance and policy that enables it, and the information sources, needs and uses that lubricate or hinder collaboration.

From this a target operating model (TOM) results. This is the desired and agreed vision that the stakeholders have of how they would like to see their CP activities improved. This captures and maps both the subjective and objective inputs of end-users on what works and what does not work in their COM of CP. From this analysis, innovation can be identified and applied across people, process, technology and governance. This analysis between current and target operating models can drive requirements, described through detailed process flows, for onward exercising, testing and validation of improvements. This is incorporated into an improvement and transformation plan to drive implementation or recommendations for changes to policy or governance.

Describing the Detailed Framework and Its Build

The TOM describes the collective improvements to CP that have been identified at community level of each pilot across the following core capability areas: Governance, People, Process and Technology. It is from the TOM that the identified short, medium- and long-term improvements are drawn, prior to their pilot testing and inclusion in a CP transformation plan.

A shared, mutually agreed TOM between stakeholders is the key output that supports the CPAF. Within which, three further outputs come together to inform the TOM creating a common baseline templates from which partner pilots and CP adaptations to each local context are derived. These are:

- CP Scenario Journey
- Current Operating Model (COM)
- Transformation Change (DELTA Map)
- Target Operating Model (TOM)

These components sit as part of the overall CP Landscape Journey seen in Fig. 8.1 and are described in the following sections.

Within each of the eight participating geographic locations, one or more specific scenarios were identified based upon local needs and context. The scenarios address a broad range of the core tasks of CP and align to the identified target minority communities, all of which were identified and defined by the research undertaken.

These scenarios provide hooks to support the six identified CP outcomes. The scenarios then formed the basis for scoping the following steps; making up the CP Landscape Journey. At each stage of the CP Landscape Journey, the scenarios were reviewed to ensure continued alignment with both the process and with the aims and objectives of the pilot to:

- Ensure mapping to the stages outlined in the CP Scenario Lifecycle
- Identifying key minority groups
- Have a theme of prevention
- Supporting the six CP outcomes

As part of the process, each pilot partner completes the CP Journey template, detailing the following criteria:

- What works well and why (Successes)?
- What does not work well (Issues)?
- Why does not it work (Underlying causes)?
- The consequences of what does not work well (Failures)
- The limitations and/or risks around Governance, People, Process and Technology (Challenges)

The stages and activities in the lifecycle provide the anchor points from which the scenario process-flows of both the COM and TOM are aligned. A baseline CP Journey template is then used to capture all core stakeholders, both organisational and individual involved in the CP scenario. This includes their tasks and processes, "what works and what does not work" in their extant approach, along with the consequences that result from those aspects that either do not work or are not performing as well as required.

Each pilot partner populates the template with detailed information extracted from their scenario, with the stakeholders identified. This is a

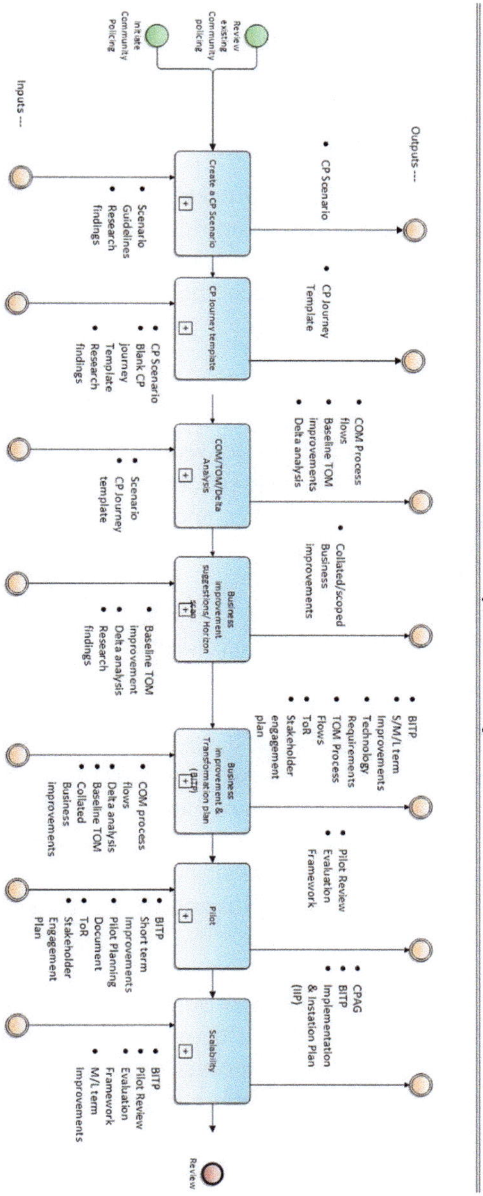

Fig. 8.1 CP landscape journey showing CPAF components (Marzell et al. 2017b)

collaborative process with all of the CP stakeholders active in the scenario; the same stakeholders remain constant and consistent, where possible, throughout the life of the pilot and its following test-beds and evaluation.

COM/TOM/Delta Analysis

From the scenario, a COM (As-Is) process flow is created by each pilot partner using a free-flow narrative. From this a desired TOM is articulated based on the delta analysis of "what works, what does not work" in the COM, that is, the difference between the COM and the TOM where improvements to CP can be made.

The COM

The COM, determined through analysis of the scenario, is then described in more detail through process flows derived from the information within the template. These are the business process activities within the CPAF that describe the step-by-step activities and tasks of each pilot partner's COM based upon the scenario provided. An example of this high-level process flow is shown in Fig. 8.2 below.

The COM process flow—or flows, as there may be several depending upon the complexity of the scenario—consists of several layers. Each layer goes into greater granularity to provide a view of the processes that support the activities that make up the current way that CP is handled in each pilot partner. This can be seen in Fig. 8.3.

This detail includes all of the tasks, functions, systems, information, governance and technology that support and enable the CP scenario to take place. Once complete, each pilot partner reviews and validates this to ensure it presents an accurate picture of how they undertake CP in the context of both their provided scenarios and their local CP context.

The TOM

From an analysis—both subjective and objective—of "what works" and "what does not work" in the COM, a desired TOM is produced. This is an identification of the high-level improvements needed to improve the areas of CP identified and to bridge the gap between the COM and the TOM. This is the delta analysis.

How these high-level improvements for the TOM might be achieved is then described, using the collective input of the (project) partners' knowledge and expertise, based upon the core capability areas of people, process,

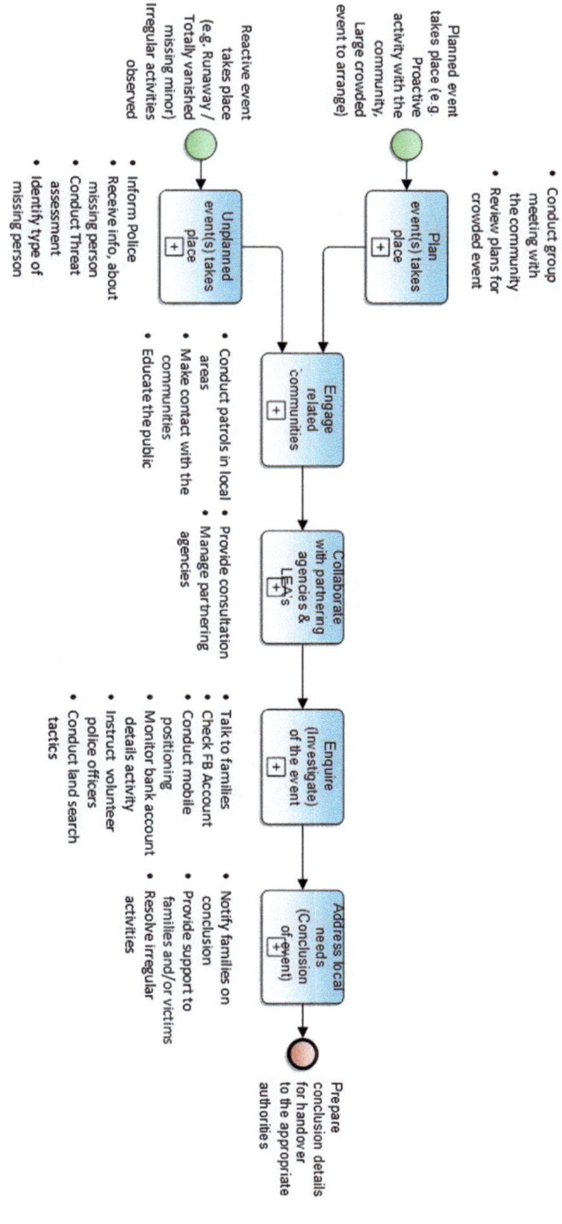

Fig. 8.2 High level process flow for a generic CP journey showing common core tasks and activities (Marzell et al. 2017b)

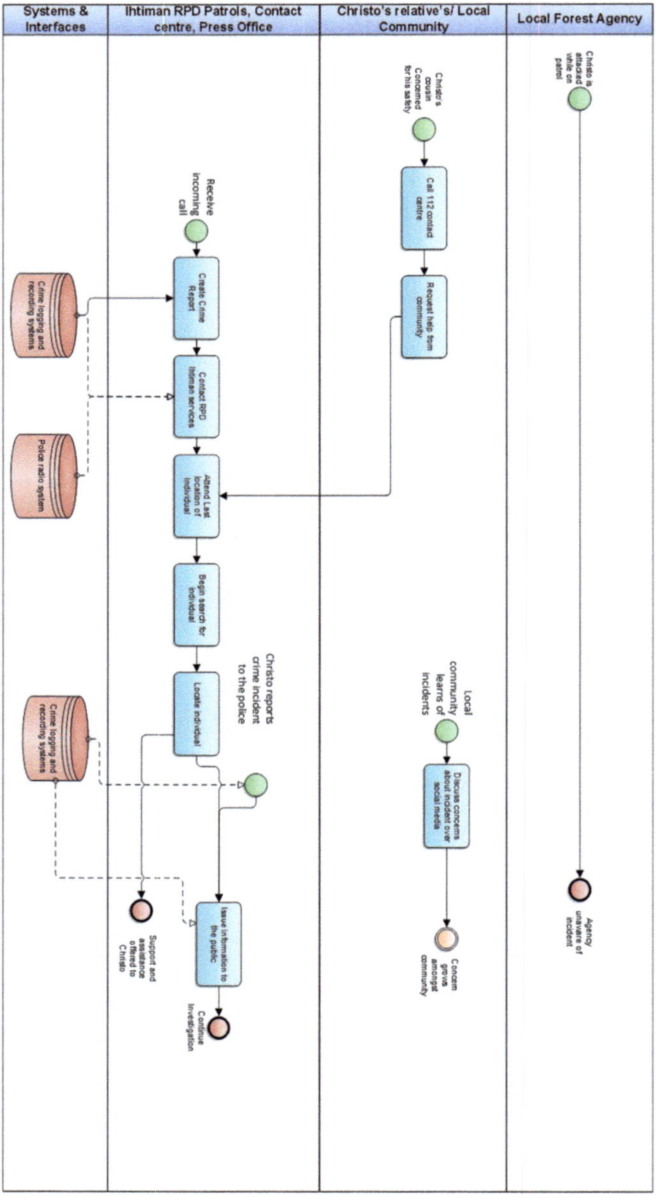

Fig. 8.3 Example of lower level process flow from a pilot COM (Marzell et al. 2017b)

governance and technology. These are the core "operational" areas that support CP activities on the ground where improvement, whether in one or several of these capability areas, need to be made to achieve the CP improvements sought.

The desired TOM, refined by the collective input of consortium partners, is given more granularity, like the COM, through the creation of detailed process flows. These represent how the desired TOM improvements to CP will be achieved, across all CP stakeholders and partners, to improve the outputs and outcomes described by the scenario. An example based upon a scenario from one of the eight research locations can be seen in Fig. 8.4.

The Delta Analysis

The principal activity that takes place in the delta analysis is an understanding of "what works" and "what does not work" within the scenarios provided and captured by the COM. This includes why aspects do not work, the consequences that result and the resultant areas of CP that need improvement.

It is a subjective and objective review from all stakeholders in the scenario, supported by the combined expertise and knowledge of the partners. Using the four core capability areas of Governance, People, Process and Systems this more detailed analysis maps to the successes, failures, challenges and issues described at each pilot partner with their scenarios through the CP COM.

It provides the basis from which an understanding of where the improvements to CP to achieve the TOM can be made. This will be either from the use or adaptation of existing CP processes and activities (either in the country or from a consortium of relevant stakeholders), wider EU best practice and research or whether new developments in people, processes, technology or governance need to take place.

An example, seen in Fig. 8.5, illustrates how the improvements to CP in the scenario identified in the delta analysis of the COM, map to the detailed process flow of the CP COM.

Horizon Scan

It is necessary that the improvements to CP identified within the desired TOM are based upon sound inputs drawn from as wide a base as possible. These identified improvements to the "business of CP", the business

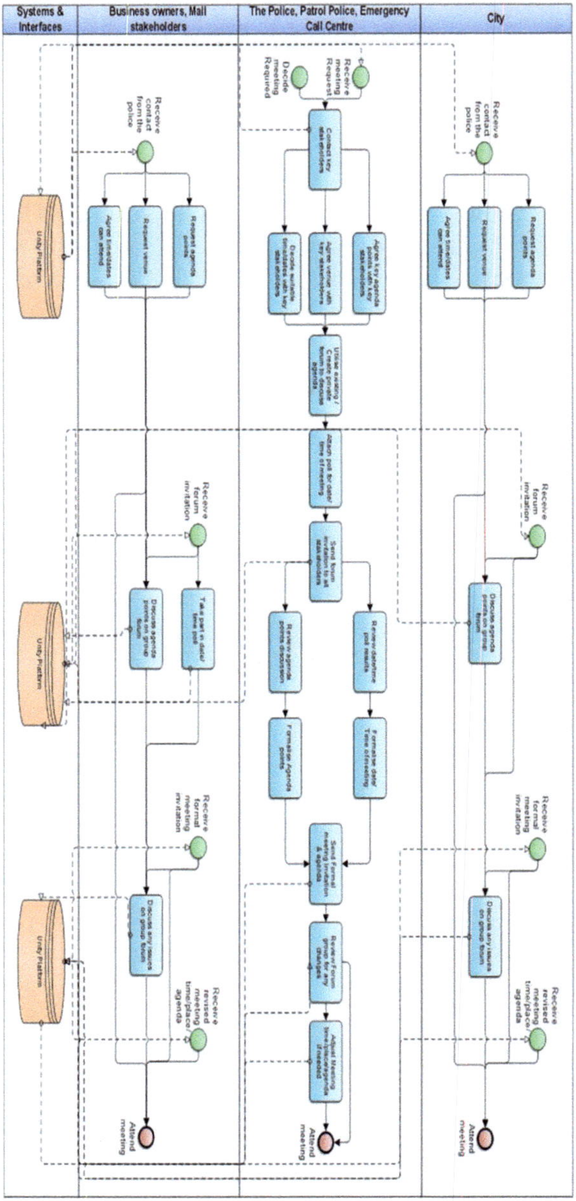

Fig. 8.4 Example of Detailed TOM process flow (Finland scenario) (Marzell et al. 2017b)

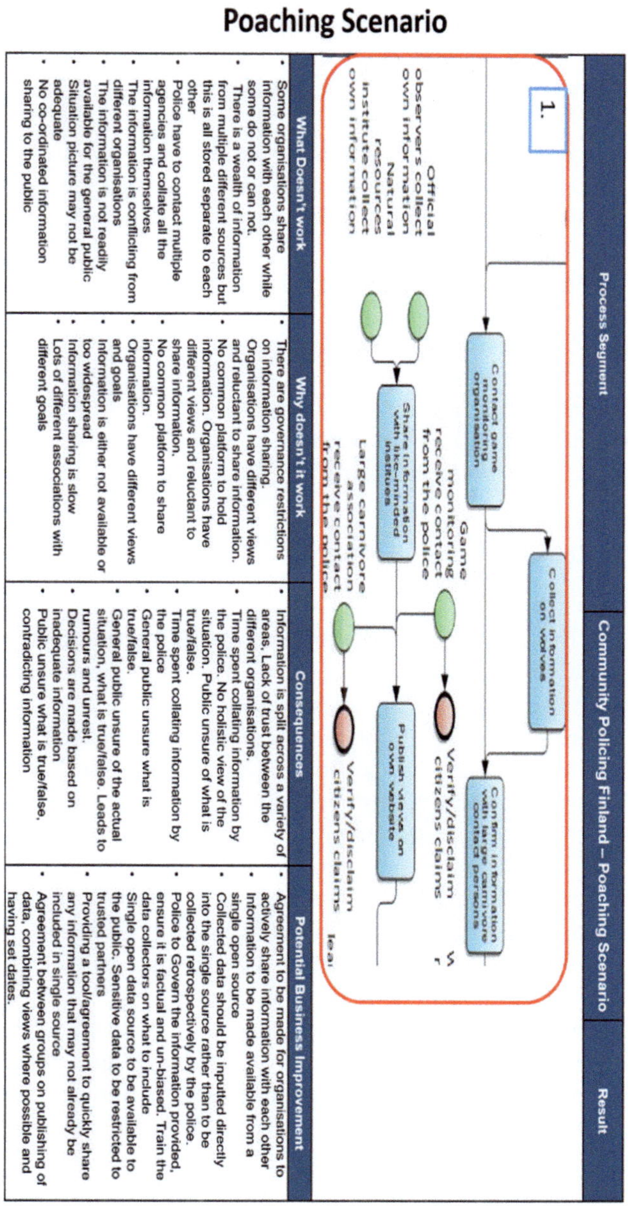

Fig. 8.5 Example of mapping the delta analysis to the COM process flows (Marzell et al. 2017b)

improvements, are subjected to a horizon scan to identify where and how these improvements can be realised based upon proven examples. This horizon scan will take account of the following:

- CP policy and practice within the pilot partner's country.
- The collective and combined knowledge and expertise from partners and their CP policy and practice, as well as direct input from relevant partners.
- The extant research on CP policy and practice across pilot partner countries.
- From EU and wider international research on CP.

The outputs of these are used to shape the TOM, how the pilot partner structures and manages its pilot and test-beds, and to inform the Business Improvement and Transformation Plan (BITP), which it overlaps.

Business Improvement and Transformation Plan (BITP)

The BITP is the culmination of the outputs from the analysis of "what works" and "what does not work" in the COM, the creation of the desired CP improvements in the TOM, and the delta analysis and horizon scan of how and where these desired improvements identified in the CP scenario might be achieved, based upon a wide, expert and empirical investigation.

The BITP will set out the business improvements to CP set against the relevant area of CP community engagement and activity, identifying where and how the improvements need to be made and whether these are in People, Process, Governance or Technology areas, or a combination of these. A ranking then takes place to determine whether these are short, medium or long-term improvements. An example BITP can be seen in Fig. 8.6.

The Transformation Plan that sits alongside the Business Improvements is the mechanism to describe how, where, who, what and when the business improvements are achieved across the short, medium and long-terms. The objective is to identify the improvements, primarily in the short term, that are most likely to achieve the desired results if tested in each pilot and follow on test-beds.

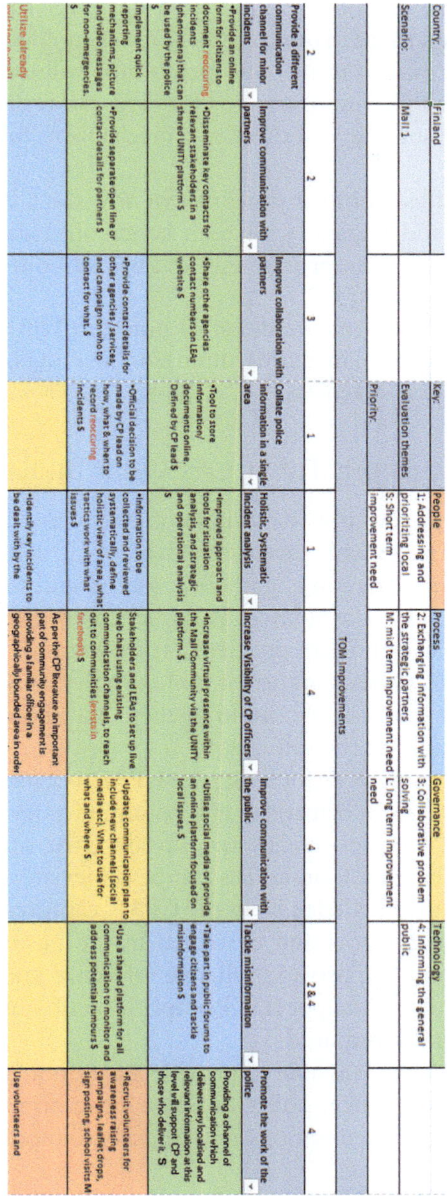

Country:	Finland		Key:	People	Process	Governance	Technology	
Scenario:	Mall 1		Evaluation themes	1: Addressing and prioritizing local	2: Exchanging information with the strategic partners	3: Collaborative problem solving	4: Informing the general public	
			Priority:	S: Short term improvement need	M: mid term improvement need	L: long term improvement need		
TOM Improvements								
2	2	3	1	1	4	4	2 & 4	4
Provide a different communication channel for minor incidents	Improve communication with partners	Improve collaboration with partners	Collate police information in a single area	Holistic, Systematic Incident analysis	Increase Visibility of CP officers	Improve communication with the public	Tackle misinformaiton	Promote the work of the police
•Provide an online form for citizens to document reoccuring incidents (phenomena) that can be used by the police S	•Disseminate key contacts for relevant stakeholders in a shared UNITY platform S	•Share other agencies contact numbers on LEAs website S	•Tool to store information/ documents online. Defined by CP lead S	•Improved approach and tools for situation analysis, and strategic and operational analysis S	•Increase virtual presence within the Mall Community via the UNITY platform. S	•Utilise social media or provide an online platform focused on local issues. S	•Take part in public forums to engage citizens and tackle misinformation S	Providing a channel of communication which delivers very locaslied and relevant information at this level will support CP and those who deliver it. S
Implement quick reporting mechanisms, picture and video messages for non-emergencies. S	•Provide separate open line or contact details for partners S	•Provide contact details for other agencies / services, and campaign on who to contact for what. S	•Official decision to be made by CP lead on how, what & when to record reoccuring incidents S	•Information to be collected and reviewed systematically, define holistic view of area, what tactics work with what issues S	Stakeholders and LEAs to set up live web chats using existing communication channels, to reach out to communities (exists in facebook) S	•Update communication plan to include new channels (social media etc). What to use for what and where. S	•Use a shared platform for all communication to monitor and address potential rumours S	•Recruit volunteers for awareness raising campaigns, leaflet drops, sign posting, school visits M
Utilize already existing e-mail				•Identify key incidents to be dealt with by the	As per the CP literature an important part of community engagement is providing a familiar officer in a geographically bounded area in order			Use volunteers and

Fig. 8.6 Example collated, categorised and prioritised BITP (Marzell et al. 2017b)

The scope of the Transformation Plan will include such mechanisms as the CP terms of reference (ToR) and the CP stakeholder engagement plan (SEP). In transformation terms, this means how will the CP stakeholders take their identified CP improvements forward and realise them in practical terms. A short summary of these two mechanisms follows:

- *ToR:* This will set out the principles of how the CP stakeholders will collaborate and engage with each other on a day-to-day basis to implement the CP improvements and to carry out meaningful community empowering CP.
- *SEP:* This will set out the detail of how the ToR will be delivered; e.g. a plan of action, including timings of meetings, roles and responsibilities, tasks and actions.

Conclusion: Trust Can Only Be Earned

Information is the common denominator between the many different organisations and stakeholder actors involved in CP at a local community level. How such information is gathered, shared and used to benefit the community and tackle the numerous challenges at the grassroots level, before they become significant problems, is paramount. Trust needs to be earned between all citizens and their communities as well as the various law enforcement and other multi-agency stakeholders involved in Community Policing on the ground.

This proposed framework for trust, the CPAF, is a mechanism to better understand and share information at a community level. It provides an essential building block to create and sustain trust and, ultimately, the means by which shared CP outcomes can be achieved.

The findings and knowledge from the CPAF, captured and described in the target operating model and described in the business improvement and transformation plan, provide valuable reference knowledge and benefits that are scalable to support CP improvement outcomes, wherever community policing is required.

Such a framework that is able to provide a common and shared understanding of the governance, policy and flow of information between the multiple stakeholder organisations involved is an important step toward establishing trust.

References

Akhgar, B., Bayerl, P. S., Markarian, G., & Karlovic, R. (2017). Concluding remarks. In *Community policing—A European perspective* (pp. 269–271). Berlin: Springer. https://doi.org/10.1007/978-3-319-53396-4.

Bayerl, P. S., van der Giessen, M., & Jacobs, G. (2015). Report on existing approaches and best/effective practices to community policing. *Unity Project.* Retrieved from https://www.unity-project.eu/wp-content/uploads/2015/12/D3.1-Report-on-Existing-Approaches-and-Best-Practices-to-CP.pdf

Brewster, B., Gibson, H., & Gunning, M. (2018). Policing the community together: The impact of technology on citizen engagement. In G. Levantakis, M. Haberfeld, & P. Papanikolaou (Eds.), *Societal implications of community-oriented policing technology.* Berlin: Springer.

Community Oriented Policing Services. (2014). *Community policing defined.* Washington, DC: Department of Justice.

Disraeli, B. (1880). *Endymion.* London: Longman.

ISO/IEC/IEEE42010:2011. (2011). *Systems and software engineering—Architecture description.* ISO Architecture.

Maguire, E. R., Kuhns, J. B., Cox, S. M., & Cox, S. M. (1997). Patterns of community policing in nonurban America. *Journal of Research in Crime and Delinquency, 34*(3), 368–394. https://doi.org/10.1177/0022427897034003004

Marzell, L., Pettengell, T., Lewis, C., & Rai, I. (2017a). First report on provision of usable community stakeholder outputs and meaningful scope for the CPAF, TOM and CONOPS. *Unity Project.* Retrieved from https://www.unity-project.eu/wp-content/uploads/2015/12/D4.5-1st-report-on-provision-of-usable-community-stakeholder-outputs-and-meaningful-scope-for-CPAF-TOM-CONOPs.pdf

Marzell, L., Pettengell, T., & Rai, I. (2017b). Third report on gap, capability and role mapping and assessment for combined effect CP initiatives between extant CP methods and CP target operating model. *Unity Project.* https://www.unity-project.eu/wp-content/uploads/2015/12/D4.7-Extant-CP-vs-TOM-3rd-Report.pdf

Peak, K. J., & Glensor, R. W. (1996). *Community policing and problem solving: Strategies and practices* (p. 436). Upper Saddle River: Pearson Prentice Hall.

Reisig, M. D., & Parks, R. B. (2004). Can community policing help the truly disadvantaged? *Crime and Delinquency, 50*(2), 139–167. https://doi.org/10.1177/0011128703253157

Renard, T., André, S., Devroe, E., Duquet, N., Lemeunier, F., Ponsaers, P., & Seron, V. (2016, October). Counterterrorism in Belgium: Key challenges and policy options. *Egmont Paper,* 89. Retrieved from http://www.egmontinstitute.be/content/uploads/2016/10/Egmont-Paper-89_OFFICIAL-FINAL.pdf?type=pdf

Rukus, J., Warner, M. E., & Zhang, X. (2018). Community policing: Least effective where need is greatest. *Crime and Delinquency, 64*(14), 1858–1881. https://doi.org/10.1177/0011128716686339

van der Giessen, M., Jacobs, G., Brein, E., & Bayerl, P. S. (2016). 1st stakeholder analysis—Comparative view on stakeholder needs and perspectives. *Unity Project.* Retrieved from https://www.unity-project.eu/wp-content/uploads/2015/12/D3.3-Comparative-View-on-Stakeholder-Needs-and-Perspectives.pdf

Vickers, L., Aston, E., Bayerl, P. S., Marzell, L., McCrone, N., Ayora, C., et al. (2018). Overall project report. *Unity project.* Retrieved from https://www.unity-project.eu/wp-content/uploads/2015/12/D1.3-Unity-Project-Report.pdf

Weine, S. (2017). *How local law enforcement uses community policing to combat terrorism.* Retrieved September 20, 2018, from https://www.lawfareblog.com/how-local-law-enforcement-uses-community-policing-combat-terrorism

PART III

Methods and ICT Approaches

CHAPTER 9

Multimedia Analysis on User-Generated Content for Safety-Oriented Applications

Nikolaos Papadakis, Antonios Litke, Anastasios Doulamis, Eftychios Protopapadakis, and Nikolaos Doulamis

Introduction

One important factor that boosts the rapid penetration of smartphone devices is the incorporation of more sensors (Saroiu and Wolman 2010). Nowadays, almost all smart mobile phones are equipped with GPS chips, accelerometers, proximity sensors and cameras. As for the latter one, most of the cameras embedded in mobile phones are of high resolution and enable image and video capturing of high quality. We can state that today

N. Papadakis
Hellenic Army Academy, Vari, Greece

National Technical University of Athens, Athens, Greece
e-mail: npapadakis@infili.com

A. Litke
Infili Technologies PC, Zografou, Greece
e-mail: alitke@infili.com

A. Doulamis (✉) · E. Protopapadakis · N. Doulamis
National Technical University of Athens, Athens, Greece
e-mail: adoulam@cs.ntua.gr; eftprot@mail.ntua.gr; ndoulam@cs.ntua.gr

B. Akhgar et al. (eds.), *Social Media Strategy in Policing*,
Security Informatics and Law Enforcement,
https://doi.org/10.1007/978-3-030-22002-0_9

considerable work is done to embed high-definition images (Agrafiotis et al. 2015; Kontogianni et al. 2015) on mobile smart devices (Gelfand et al. 2010). These high-resolution sensors have stimulated a new type of content, the so-called *user-generated content* (Krumm et al. 2008; Soursos and Doulamis 2012).

The advent of user-generated content (UGC) has reshaped the online video market enormously (Cha et al. 2009). Nowadays, hundreds of millions of Internet users are not only content consumers but also publishers. The duration of UGC videos is shorter by two orders of magnitude than traditional videos and so is their production time. Wired magazine refers to this small-sized content pop culture as "bite-sized bits for high-speed munching" (Cha et al. 2009). This trend is also popular with text messages like Twitter. Nowadays, many people, even politicians, are chatting using tweets. Indeed, some of these messages are now used for 3D cultural heritage reconstruction in the wild (Makantasis et al. 2016), using images from the Web (Doulamis and Katsaros 2016).

The need for applying algorithms that enable the detection of events over social media is critical (Doulamis et al. 2016; Kokkinos et al. 2013; Popescu and Pennacchiotti 2010; Weng and Lee 2011). The current advancement of surveillance systems and early warning architectures is crucial for securing humans in large cities and prevent illegal actions (Addington 2009; Fujii et al. 2010). This proves to be a key milestone to European economies, industries, authorities and, at the end, citizens. Nevertheless, the cost for large deployments and maintenance of ground sensing networks for local surveillance is extremely high, especially when the surveillance is operated as a manual environment (Sardis et al. 2010). We, as humans, are subjective in interpreting risks and may concentrate our attention on different objects rather than on the real threat; most importantly we easily get tired, especially in cases where we are obliged to monitor systems with multiple cameras for the chance of detecting trespassing or a citizen's improper behaviour (Doulamis and Doulamis 2012).

Related Work

The recent advantages in hardware and software technologies have significantly reduced the cost of visual sensors, and they have forced the development of new innovative automated (or even semi-automated) algorithms (Grabner et al. 2008; Rosenberg et al. 2005) that are able to localize image objects/regions of interest so as to alert for potential risks

(Kosmopoulos et al. 2012; Voulodimos et al. 2012). Nowadays, there exist several surveillance cameras in cities not only for security and preventing crimes but also for traffic management and control (Coleman 2004; Fyfe and Bannister 1996; Koskela 2000). On the other hand, the Internet, new digital technologies and social media services like Facebook and Twitter are transforming our world in every walk of life and in every line of business. Nowadays, several trillions of images exist in loosely structured repositories (the Web, file servers, databases, etc.), and their number grows rapidly every day (Doulamis et al. 2003a, b). This is a natural outcome for a number of reasons such as the low-cost of digital cameras, low-cost storage and easy Web hosting and, on top of that, the need of people and organizations to share multimedia files, either for social or commercial purposes. This huge amount of media information presents an opportunity for the computer vision and multimedia society to "perform automatic or semi-automatic" analysis of the content and localize objects of interest (Doulamis et al. 1999a, 2000a, 2001).

The key problem in such scenarios is, however, that user-generated content cannot be trusted and contains noise (Lenders et al. 2008). Traditionally, a news article published, for example, on a well-known website like Reuters or BBC carried some implicit quality and authenticity guarantees based on the reputation of the news provider. In contrast, content from citizen journalists may come from multiple unknown individuals and may be published on possibly untrusted sites, which do not naturally embody similar a priori confidence in the contents (Borden and Tew 2007; Lenders et al. 2008).

Our Contribution

The key goal of this chapter is to incorporate image analysis methods into applications that provide a more safe and secure environment for citizens in large-scale cities. Towards this end, we need to process, evaluate and prioritize video and audio streams generated by users (where users act as "sensors") in a participatory urbanism approach. The proposed method prioritizes multimedia streams based on their content (e.g. streams that involve critical situations such as panic of crowds, ongoing street riots or other dangerous situations), thus enabling authorities to better shortlist the actual relevant streams and identify critical events.

To achieve this, a set of visual features is extracted. Ideally, such a set should identify the content of any image as well as being possible in *real time*. Thus, we need to identify low-level features. One example that can be used is the AC component of the discrete cosine transform (DCT) that expresses the texture properties of an image. An image with high texture contains more complicated and cluttered content than other types of content in which more uniform regions are encountered. This descriptor was used in recent years for the unsupervised tracking and classification of video streams (Doulamis et al. 2000a, 2003a, b; Ntalianis et al. 2015) and for content-based image retrieval either from 2D videos or from stereoscopic 2.5D or 3D image sequences (Doulamis et al. 2000a, b, c, d).

The main advantage of such descriptions is that the AC coefficients of the DCT are directly available in the MPEG sequence and thus no further decompression is required. Hence, this algorithm guarantees that the analysis is carried out in real time. This has been proven in one of our previous works (Akrivas et al. 2000). The motion information, available in the MPEG sequence, is also exploited to detect temporal changes in the image sequence.

The generated feature values describe various areas on the image, that is, non-overlapping image blocks. The similarity of the blocks is then assessed, and a number of clusters are created. A further investigation in clusters' number fluctuations over corresponding video frames can indicate significant changes in the content.

This chapter is organized as follows: In section "Related Work", the proposed architecture is presented. Section "Our Contribution" describes the feature extraction methodology, while section "Proposed Architecture" presents the prioritization scheme for the video streams. Experimental results are given in section "Experimental Results", and section "Conclusion" presents the conclusions of this chapter.

Proposed Architecture

Figure 9.1 shows the architecture of the proposed scheme. In a scene of interest within a city, for instance, when a security threat is noticed, several people can capture the event taking pictures and uploading this rich amount of content onto a dedicated data server. This server will stimulate responders that should take imminent actions for safety and security purposes.

The main problems in such a situation are that (1) the number of potential users who will wish to upload their content on the server will often be large and (2) the users are un-trusted in the sense that they can

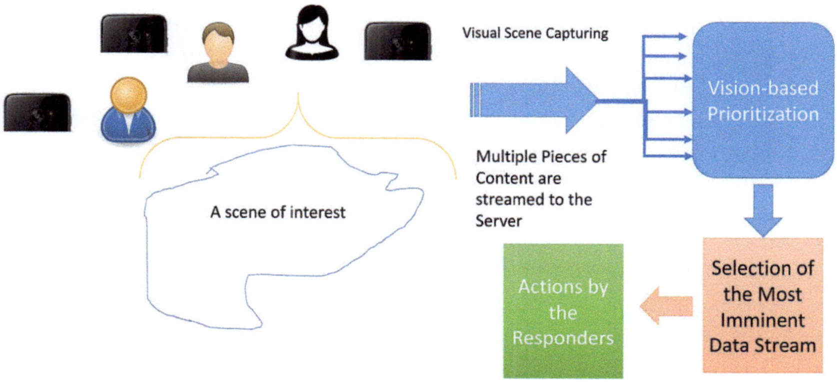

Fig. 9.1 A block diagram of the proposed architecture

stream content, which is not of high interest (priority), or that they can capture content views of the scene, which are not crucial for the responders.

The time needed to select which is the content of the highest priority is a crucial matter, which is addressed in this chapter. More specially, a vision-based prioritization toolkit is proposed, so that the streamed videos are examined with respect to their content. In order to accomplish this, we apply computer vision algorithms and machine learning techniques that can "see" inside the data. Then, the content is analyzed and the scenes that might have critical content are given higher priority than the rest.

Feature Extraction

The video sequences are first split into a set of shots. This allows for a semantic representation of the visual content. We exploit content of the MPEG sequence to accelerate the processing time, especially in mobile applications, where many users submit their queries to the system.

Scene Detection

Since a video sequence is a collection of different shots, each of which corresponds to a continuous action of a single camera operation, a shot detection algorithm is first applied in order to temporally segment the sequence into shots. In our approach the DC coefficients of the DCT transform is exploited as in Akrivas et al. (2000). These coefficients are directly available in the case of intracoded frames (I frames) of MPEG compressed video sequences, while for the intercoded ones (P and B frames) they can be estimated by the motion compensated error with minimal decoding effort. The adoption of the above-described algorithm results in a significant reduction of the required computations compared to other algorithms, which require a full resolution search of the video data.

Extracting Features in the Compressed Domain

The MPEG compressed video streams consist of three types of frames: Intracoded (I), Predictive (P) and Interpolative (B) frames (Doulamis et al. 1999a, b; Pancha and Zarki 1993). The I frames are derived by the application of an 8 x 8 DCT on the original image data followed by a quantization matrix using run length coding.

In P frames, macroblocks are usually coded with the application of a motion estimation algorithm. Motion vectors are estimated by searching a square area around each macroblock of the previous I or P frame in order to find a motion vector that minimizes the absolute difference between the current macroblock and the chosen one in the previous frame. If the absolute difference is less than a predefined threshold, the motion vector is coded and transmitted. Then a DCT transform is applied to the prediction error of each macroblock.

B frames coding is similar to the procedure described for P frames. The only difference is that the motion vector can be estimated with respect to the previous I, P frames or the following I, P frames or an interpolation between them.

These three types of frames are transmitted using a predefined pattern, which is called *Group of Picture* (GoP). This pattern remains the same for all frames of the same sequence and is embedded in the compressed stream in the form of overhead headings so that the decoder can localize the I, P and B frames.

Spatial Description

Once a video sequence is temporally partitioned into video shots, the next step of the analysis is to classify the areas. Towards this, we use the AC coefficients of the DCT transform on a region of 8 x 8 pixels. This patch is selected, since it is compatible with the MPEG standard, so that no decoding is required (Brandenburg and Stoll 1994).

The features extracted are totally in alignment with the MPEG compressed stream, so that no decoding of the streams is required. This allows for a real-time implementation of the analysis without delaying the decompression process.

Let us denote as R_i the 8 × 8 patch of the image. Let us also assume that we have N patches available. Then, the DCT transform on the image derive a new one-to-one region, called F_i, that contains the spatial frequencies of the pixel content onto the x/y-direction.

$$F_i = \mathrm{DCT}\left(R_i\right)_{8\times 8} \quad \forall i = 1,\ldots,N \tag{9.1}$$

where function DCT(•) is the DCT transform of the image region R_i and F_i is of the same size as the image patch, that is, 8 × 8 pixels.

In the following, we vectorize F_i to create a vector of 64 × 1 elements; then we drop the first element, which correspondents to the DC coefficient, that is, the average illumination within the patch. More specifically, let us denote as $f_i = \mathrm{vec}(F_i)$ the vectorized form of F_i. The derived AC coefficients of F, called AC_i, are

$$\mathrm{AC}_i = f_{i+1}, \quad \forall i = 1,\ldots,N-1 \tag{9.2}$$

with f_j being the jth element of the vector f. Equation (9.2) is valid, since we remove the first element from f_i, that is, the DC coefficient.

Gathering all values for all patches N, we can form a vector AC for each patch as $\mathrm{AC} = [\mathrm{AC}_i]$. This vector can be extended for all N available patches.

Temporal Description

The next step is to extract information regarding the motion of the scene. This is achieved through the exploitation of the motion vectors that indicate the direction of the motion on objects of interest. Motion information is extracted using again larger regions of 16 × 16 pixels, which are in fact four 8 × 8 pixel regions. In this way, the local activity of the scene is described.

Let us denote as $\mathrm{mv}_i = (v_{x,i}, v_{y,i})_{16\times 16}$ the motion vector onto an area of 16 × 16 pixels. This vector describes the motion direction in the image with respect to the two spatial directions of the image. We also denote as $\mathrm{mv}_i = [(v_{x,i}, v_{y,i})]^T$. The total image vector is denoted as $g = [\mathrm{AC}\ \mathrm{mv}\ ]^T$.

Fuzzy Representation

Since visual noise is often part of the captured data, we fuzzify the features so that they can belong to more than one class. In this way, the potential noise is reduced, and the visual representation is improved. Fuzzy representation has been adopted in other visual analysis applications such as the ones proposed in Doulamis and Doulamis (2001) and Vertan and Boujemaa (2000).

In particular, let us denote as $\mu(\bullet)$ the fuzzy membership functions. In this chapter, the simplest functions such as the triangular ones have been adopted. The features vectors used will then be

$$g_f = \mu(g) \tag{9.3}$$

where vector g_f is the fuzzy version of the feature.

Video Stream Prioritization

In the previous section, we have extracted a set of features per video frame to represent its visual complexity and motion. In this section, we introduce a machine learning-based algorithm that enables the identification of the most important video shots from a pool of multiple video streams. In this way, we prioritize the streams by “looking inside the data”, so that streams that do not depict similar visual content are given more attention than other scenes. We assume that these scenes contain “abnormal behaviour” material and can thus be considered more important by the responders.

For the selection of the most important frames, we apply a clustering algorithm. The goal of this approach is to group frames into visually meaningful clusters, taking into account the diversity of the content. Then, we extract the key frame or key frames within each cluster, which are shown to the responders to take actions. The algorithm gives higher prioritization scores to the key frames rather than to remaining rest frames of the transmitted video streams.

Density-Based Clustering as a Basis for Profile Modelling

In this chapter, a density-based clustering algorithm is introduced. Density-based schemes are extensively used for user profile modelling (Karamolegkos et al. 2009) or image clustering (Makantasis et al. 2016). In density-based clustering, a cluster is a set of data objects spread in the data space over a contiguous region of high density objects. Density-based clusters are separated from each other by contiguous regions of low density of objects.

The algorithm applied works at two discrete phases. The first localizes the data with respect to their density properties on the feature space. This is done by applying the OPTICS approach (Ankerst et al. 1999). OPTICS is based on DBSCAN (Ester et al. 1996) and the work in Stuetzle (2003).

OPTICS computes a minimum spanning tree of the data, where edge weights represent pairwise distances. These distances are smoothed by a density estimator, called core distance. The core distance of a point x_i is the smallest threshold r such that x_i is still considered a core object by the DBSCAN algorithm, that is, x_i has still at least k objects in its neighbourhood with radius r. The resulting distance, which is used to construct the Minimum Spanning Tree, is called reachability distance (RD). Taking k as input parameter for smoothing the density estimation, the reachability distance of point x_i is defined relative to a reference object y as the minimum of the core distance of y and the actual distance between x_i and y.

Then, the second phase estimates the boundaries of the clusters using a plain signal process, which estimates the local maxima (i.e. peaks) over the waveform generated by the RDs according to the data assessment order. Valleys among the peaks are considered clusters. We select this clustering approach, since it works without knowing the exact number of clusters a priori something, which is crucial for our application scenarios as the complexity of the visual streams gives the priorities on the input data.

Experimental Results

The evaluation process was based on YouTube videos using the following tags: "car explosion" and "dash camera". The datasets contain all types of variance sources including: (1) different capturing angles, (2) movement speed, (3) occlusions (i.e. other vehicles or pedestrians) and (4) explosion types and illuminance changes.

Videos are processed frame by frame. Each frame is divided into a constant number of non-overlapping image patches. A set of feature descriptors is extracted for each patch. Thus, each frame is described by matrix $D \in R^{m \times n}$, where m denotes the number of patches and n stands for the number of feature values. Given the D matrix and the RDs (from OPTICS), we estimate the number of clusters in each frame.

In our case, an anomaly can be described as any event that causes either a significant change in the number of clusters between two successive frames or a greater variance in cluster numbers for consecutive frames, as illustrated in Fig. 9.2.

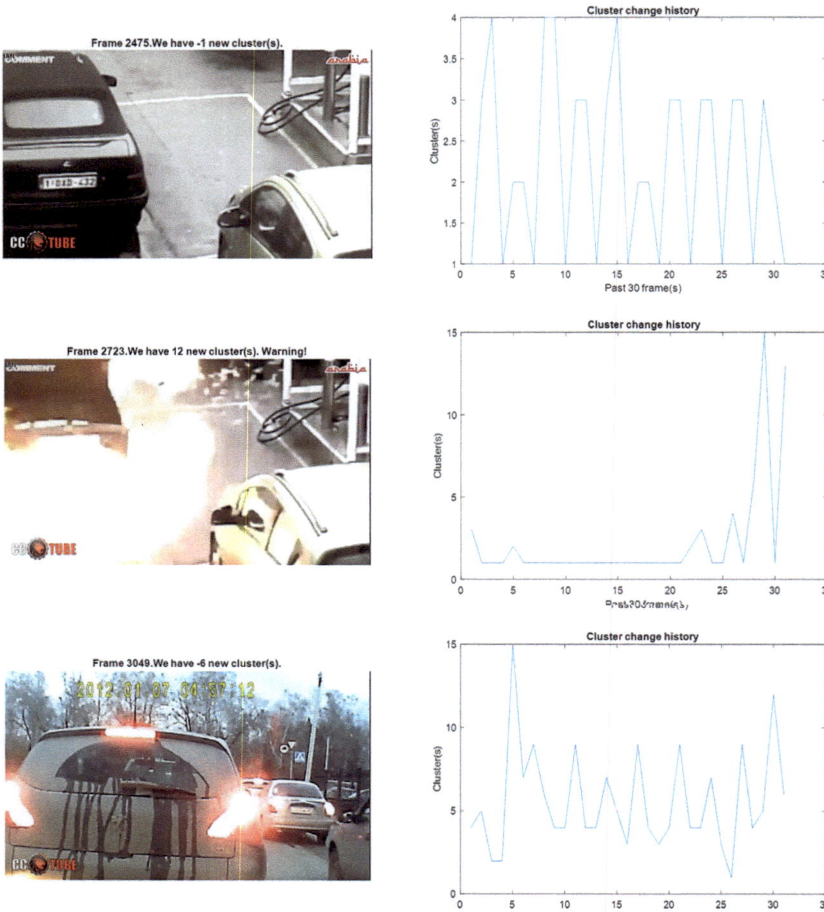

Fig. 9.2 Illustration of the proposed anomaly detection scheme. The mechanism is triggered by either a significant change in the number of clusters or by variance differentiation

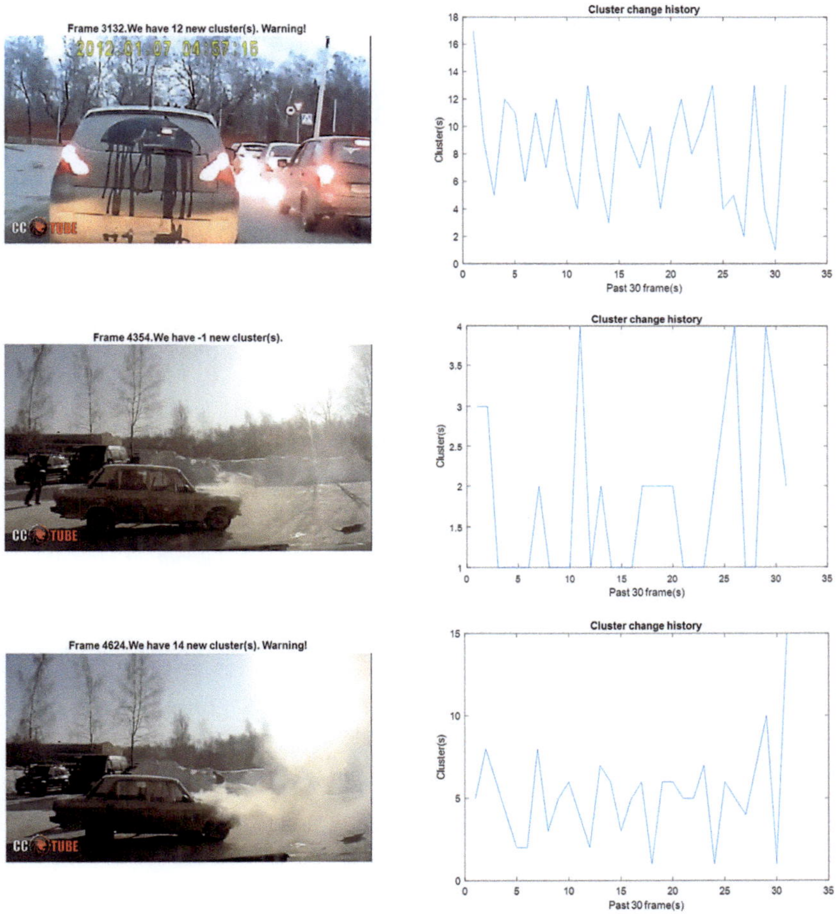

Fig. 9.2 (continued)

Conclusions

In this chapter, we present an anomaly detection approach in videos. The proposed methodology makes no assumptions regarding the application scenario and video quality. Frames are compared to each other by clustering the image patches. More sophisticated signal analysis techniques could be employed in order to strengthen detection performance.

References

Addington, L. A. (2009). Cops and cameras: Public school security as a policy response to columbine. *The American Behavioral Scientist, 52*, 1426–1446. https://doi.org/10.1177/0002764209332556

Agrafiotis, P., Stathopoulou, E. K., Georgopoulos, A., & Doulamis, A. D. (2015). HDR imaging for enhancing people detection and tracking in indoor environments. In *VISAPP* (Vol. 2, pp. 623–630).

Akrivas, G., Doulamis, N. D., Doulamis, A. D., & Kollias, D. (2000). *Scene detection methods for MPEG-encoded video signals.* Presented at the 2000 10th Mediterranean Electrotechnical Conference. Information Technology and Electrotechnology for the Mediterranean Countries. Proceedings. MeleCon 2000 (Cat. No.00CH37099) (Vol. 2, pp. 677–680).. https://doi.org/10.1109/MELCON.2000.880024.

Ankerst, M., Breunig, M. M., Kriegel, H. -P., & Sander, J. (1999). *OPTICS: Ordering points to identify the clustering structure.* Proceedings of the 1999 ACM SIGMOD International Conference on Management of Data, SIGMOD '99 (pp. 49–60). New York: ACM. https://doi.org/10.1145/304182.304187.

Borden, S. L., & Tew, C. (2007). The role of journalist and the performance of journalism: Ethical lessons from "fake" news (seriously). *Journal of Mass Media Ethics, 22*, 300–314. https://doi.org/10.1080/08900520701583586

Brandenburg, K., & Stoll, G. (1994). ISO/MPEG-1 audio: A generic standard for coding of high-quality digital audio. *Journal of the Audio Engineering Society, 42*, 780–792.

Cha, M., Kwak, H., Rodriguez, P., Ahn, Y.-Y., & Moon, S. (2009). Analyzing the video popularity characteristics of large-scale user generated content systems. *IEEE ACM Transactions on Network, 17*, 1357–1370. https://doi.org/10.1109/TNET.2008.2011358

Coleman, R. (2004). Images from a Neoliberal City: The state, surveillance and social control. *Critical Criminology, 12*, 21–42. https://doi.org/10.1023/B:CRIT.0000024443.08828.d8

Doulamis, A., & Doulamis, N. (2001). *Fuzzy histograms for efficient visual content representation: Application to content-based image retrieval.* IEEE International Conference on Multimedia and Expo. ICME 2001.(ICME) (p. 227). https://doi.org/10.1109/ICME.2001.1237866.

Doulamis, N., & Doulamis, A. (2012). Fast and adaptive deep fusion learning for detecting visual objects. In A. Fusiello, V. Murino, & R. Cucchiara (Eds.), *Computer vision—ECCV 2012. Workshops and demonstrations, lecture notes in computer science* (pp. 345–354). Berlin: Springer.

Doulamis, A., & Katsaros, G. (2016). *3D modelling of cultural heritage objects from photos posted over the Twitter.* Presented at the 2016 IEEE International Conference on Imaging Systems and Techniques (IST) (pp. 389–394). https://doi.org/10.1109/IST.2016.7738257.

Doulamis, N. D., Doulamis, A. D., Avrithis, Y., & Kollias, S. D. (1999a). *A stochastic framework for optimal key frame extraction from MPEG video databases.* Presented at the 1999 IEEE Third Workshop on Multimedia Signal Processing (Cat. No.99TH8451) (pp. 141–146). https://doi.org/10.1109/MMSP.1999.793811.

Doulamis, A. D., Doulamis, N. D., Ntalianis, K. S., & Kollias, S. D. (1999b). *Unsupervised semantic object segmentation of stereoscopic video sequences.* Presented at the Proceedings 1999 International Conference on Information Intelligence and Systems (Cat. No.PR00446) (pp. 527–533). https://doi.org/10.1109/ICIIS.1999.810342.

Doulamis, N. D., Doulamis, A. D., Avrithis, Y. S., Ntalianis, K. S., & Kollias, S. D. (2000a). Efficient summarization of stereoscopic video sequences. *IEEE Transactions on Circuits and Systems for Video Technology, 10*, 501–517. https://doi.org/10.1109/76.844996

Doulamis, A. D., Doulamis, N. D., & Kollias, S. D. (2000b). On-line retrainable neural networks: Improving the performance of neural networks in image analysis problems. *IEEE Transactions on Neural Networks, 11*, 137–155. https://doi.org/10.1109/72.822517

Doulamis, A. D., Doulamis, N. D., & Kollias, S. D. (2000c). A fuzzy video content representation for video summarization and content-based retrieval. *Signal Processing, 80*, 1049–1067. https://doi.org/10.1016/S0165-1684(00)00019-0

Doulamis, A., Doulamis, N., Ntalianis, K., & Kollias, S. (2000d). Efficient unsupervised content-based segmentation in stereoscopic video sequences. *International Journal on Artificial Intelligence Tools, 09*, 277–303. https://doi.org/10.1142/S0218213000000197

Doulamis, A., Doulamis, N., & Maragos, P. (2001). *Generalized multiscale connected operators with applications to granulometric image analysis.* Presented at the Proceedings 2001 International Conference on Image Processing (Cat. No.01CH37205) (Vol. 3, pp. 684–687). https://doi.org/10.1109/ICIP.2001.958211

Doulamis, A., Doulamis, N., Ntalianis, K., & Kollias, S. (2003a). An efficient fully unsupervised video object segmentation scheme using an adaptive neural-network classifier architecture. *IEEE Transactions on Neural Networks, 14*, 616–630. https://doi.org/10.1109/TNN.2003.810605

Doulamis, N. D., Doulamis, A. D., & Varvarigou, T. A. (2003b). Adaptive algorithms for interactive multimedia. *IEEE Multimedia, 10*, 38–47. https://doi.org/10.1109/MMUL.2003.1237549

Doulamis, N. D., Doulamis, A. D., Kokkinos, P., & Varvarigos, E. M. (2016). Event detection in twitter microblogging. *IEEE Transactions on Cybernetics, 46*, 2810–2824. https://doi.org/10.1109/TCYB.2015.2489841

Ester, M., Kriegel, H. P., Sander, J., & Xu, X. (1996). *A density-based algorithm for discovering clusters in large spatial databases with noise. In Kdd, 96(34)*, 226–231).

Fujii, Y., Maru, K., Kobayashi, K., Yoshiura, N., Ohta, N., Ueda, H., et al. (2010). e-JIKEI Network using e-JIKEI Cameras: Community security using considerable number of cheap stand-alone cameras. *Safety Science, 48*, 921–925. https://doi.org/10.1016/j.ssci.2010.03.018

Fyfe, N. R., & Bannister, J. (1996). City watching: Closed circuit television surveillance in public spaces. *Area, 28*, 37–46.

Gelfand, N., Adams, A., Park, S. H., & Pulli, K. (2010). *Multi-exposure imaging on mobile devices.* Proceedings of the 18th ACM International Conference on Multimedia, MM '10 (pp. 823–826). New York: ACM. https://doi.org/10.1145/1873951.1874088.

Grabner, H., Leistner, C., & Bischof, H. (2008). Semi-supervised on-line boosting for robust tracking. In D. Forsyth, P. Torr, & A. Zisserman (Eds.), *Computer vision—ECCV 2008, Lecture Notes in Computer Science* (pp. 234–247). Berlin: Springer.

Karamolegkos, P. N., Patrikakis, C. Z., Doulamis, N. D., Vlacheas, P. T., & Nikolakopoulos, I. G. (2009). An evaluation study of clustering algorithms in the scope of user communities assessment. *Computers & Mathematcs with Applications, 58*, 1498–1519. https://doi.org/10.1016/j.camwa.2009.05.023

Kokkinos, P. C., Koumoutsos, K., Doulamis, N. D., Varvarigos, E. A., Petrantonakis, D., Kardara, M., Sardis, E., Vekris, A., & Gerontidis, S. (2013). PERIKLIS-electronic democracy in the 21st century using mobile and social network technologies. In *EGOV/EPart ongoing research* (pp. 242–249). Citeseer.

Kontogianni, G., Stathopoulou, E. K., Georgopoulos, A., & Doulamis, A. (2015). HDR imaging for feature detection on detailed architectural scenes. *International Archives of the Photogrammetry, Remote Sensing and Spatial Information Sciences, XL-5/W4*, 325–330.

Koskela, H. (2000). 'The gaze without eyes': Video-surveillance and the changing nature of urban space. *Progress in Human Geography, 24*, 243–265. https://doi.org/10.1191/030913200668791096

Kosmopoulos, D. I., Doulamis, N. D., & Voulodimos, A. S. (2012). Bayesian filter based behavior recognition in workflows allowing for user feedback. *Computer Vision and Image Understanding, 116*, 422–434.

Krumm, J., Davies, N., & Narayanaswami, C. (2008). User-generated content. *IEEE Pervasive Computing, 7*, 10–11. https://doi.org/10.1109/MPRV.2008.85

Lenders, V., Koukoumidis, E., Zhang, P., & Martonosi, M. (2008). *Location-based trust for mobile user-generated content: Applications, challenges and implementations.* Proceedings of the 9th workshop on mobile computing systems and applications, HotMobile '08 (pp. 60–64). New York: ACM. https://doi.org/10.1145/1411759.1411775.

Makantasis, K., Doulamis, A., Doulamis, N., & Ioannides, M. (2016). In the wild image retrieval and clustering for 3D cultural heritage landmarks reconstruction. *Multimedia Tools and Applications, 75*, 3593–3629. https://doi.org/10.1007/s11042-014-2191-z

Ntalianis, K. S., Doulamis, A. D., Doulamis, N. D., Mastorakis, N. E., & Drigas, A. S. (2015). Unsupervised segmentation of stereoscopic video objects: Constrained segmentation fusion versus greedy active contours. *Journal of Signal Processing Systems, 81*, 153–181. https://doi.org/10.1007/s11265-014-0921-0

Pancha, P., & Zarki, M. E. (1993). Bandwidth-allocation schemes for variable-bit-rate MPEG sources in ATM networks. *IEEE Transactions on Circuits and Systems for Video Technology, 3*, 190–198. https://doi.org/10.1109/76.224229

Popescu, A. -M., & Pennacchiotti, M. (2010). *Detecting controversial events from twitter*. Proceedings of the 19th ACM International Conference on Information and Knowledge Management, CIKM '10 (pp. 1873–1876). New York: ACM. https://doi.org/10.1145/1871437.1871751.

Rosenberg, C., Hebert, M., & Schneiderman, H. (2005). *Semi-supervised self-training of object detection models*. WACV/MOTION (pp. 29–36).

Sardis, E., Voulodimos, A., Anagnostopoulos, V., Lalos, C., Doulamis, A., & Kosmopoulos, D. (2010). *An industrial video surveillance system for quality assurance of a manufactory assembly*. Proceedings of the 3rd International Conference on PErvasive Technologies Related to Assistive Environments, PETRA '10 (pp. 66:1–66:2). New York: ACM. https://doi.org/10.1145/1839294.1839373.

Saroiu, S., & Wolman, A. (2010). *I am a sensor, and I approve this message*. Proceedings of the 11th Workshop on Mobile Computing Systems & Applications, HotMobile '10 (pp. 37–42). New York: ACM. https://doi.org/10.1145/1734583.1734593.

Soursos, S., & Doulamis, N. (2012). *Connected TV and beyond*. Consumer communications and networking conference (CCNC), 2012 IEEE (pp. 582–586). IEEE.

Stuetzle, W. (2003). Estimating the cluster tree of a density by analyzing the minimal spanning tree of a sample. *Journal of Classification, 20*, 025–047. https://doi.org/10.1007/s00357-003-0004-6

Vertan, C., & Boujemaa, N. (2000). Using fuzzy histograms and distances for color image retrieval. *In Challenge of image retrieval retrieval, 6*, 1–6

Voulodimos, A. S., Doulamis, N. D., Kosmopoulos, D. I., & Varvarigou, T. A. (2012). Improving multi-camera activity recognition by employing neural network based readjustment. *Applied Artificial Intelligence, 26*, 97–118. https://doi.org/10.1080/08839514.2012.629540

Weng, J., & Lee, B.-S. (2011). Event detection in twitter. *ICWSM, 11*, 401–408.

CHAPTER 10

Suggesting a Hybrid Approach: Mobile Apps with Big Data Analysis to Report and Prevent Crimes

Abdi Fidow, Ahmed Hassan, Mahamed Iman, X. Cheng, M. Petridis, and Clifford Sule

Introduction

The various forms of antisocial behaviour said to be blighting particular areas of Greater Manchester, UK, have forced the local constabulary into taking tougher actions, in relation to both the miscreants involved and their parents or guardians. A new initiative started in Manchester, which is predicated on bringing back "bobby on the beat" style policing and has impressed residents in the areas of Tameside South, where police have also begun occupying "antisocial behaviour buses" and patrolling troubled crime hotspots. Under this new initiative, police will henceforth be patrolling the Hyde, Hattersley and Hollingworth areas of Greater Manchester with a view to clamping down on youth misdemeanours (RT International 2017).

A. Fidow (✉) · A. Hassan · M. Iman · X. Cheng · M. Petridis · C. Sule
Middlesex University, London, UK
e-mail: af767@live.mdx.ac.uk; ah1574@live.mdx.ac.uk; jg752@live.mdx.ac.uk; x.cheng@mdx.ac.uk; m.petridis@mdx.ac.uk; c.sule@mdx.ac.uk

B. Akhgar et al. (eds.), *Social Media Strategy in Policing*,
Security Informatics and Law Enforcement,
https://doi.org/10.1007/978-3-030-22002-0_10

This "hands-on" interventionist approach favoured by Greater Manchester police has not been universally endorsed elsewhere. In Sussex, for example, the local Chief Constable, Giles York, is advocating that it would be more sensible and economical to utilise online procedures for tackling these forms of crime. According to Mr. York, there is "*no added value*" in visiting victims of minor crimes. He suggested that it would be more "*convenient*" for members of the public to interact with the police online. His justification for this approach is that in our digital world today "*going out and meeting in the real world is sometimes a rarity*" as more is done online (Hall 2017). Therefore, why not embrace and apply it to policing too? York said businesses should also report crimes online as "it is a lot more convenient for them to fill in their own form, send documents and CCTV" (Hall 2017). He said many police forces now allow the public to report crimes via either social media or online, and in the future, victims will be able to track the progress of their investigations through a secure website (Hall 2017). This is where a smart app like the Hybrid Smart Crime Reporting App (HIVICRA) we propose in this chapter can help police and crime analysts. HIVICRA is an infographic visual intelligent crime reporting analysis app that incorporates crime data sourced from any local police, social media and crowdsourcing. It is another resource, another set of eyes on the streets for the police in their fight against crime.

Many studies about the effectiveness of community policing have been conducted over the years such as Community Crime Prevention in high-crime areas (e.g. the Seattle Neighbourhood Group Hot Spots Project by Telep and Hibdon 2018). Similarly, Meyer and Graan (2011) in their case study about the effectiveness of community policing practices in the South African Roodekrans Neighbourhood Watch West Rand illustrated the benefits of empowering communities to take ownership of their policing. Therefore, the objective of this chapter is not to repeat these studies but to take advantage of some of the research that has proven effective in crime reduction and prevention. One of the core principles of community policing according to John Alderson is "*policing with the consent and support of the community*", hence, playing the role of "*social as opposed to legal action*" (University of Portsmouth 2013). What this means is that flexibility and developing personal relationships with the communities at beat level is crucial (Ferreira 1996; cp. Chap. 4).

Numerous efforts have been made to utilise diverse contents, opinions and social behaviours from microblogs to predicting real-time notifications for social or public health risks using analysis of phrases expressing sentiments (Wang et al. 2017). There are various types of private and open source Internet sites such as Facebook, Instagram and blogs that Internet users post, share and view contents about themselves and their friends ranging from their movements, thoughts as well as their likes and dislikes (Omand et al. 2014). All of these interactions and shared contents leave digital traces, which can be a source of valuable information for law enforcement agencies with the right capabilities. However, until recently insufficient attention has been paid to how user-generated social media contents can be used for crime prediction and their effects, for example through Twitter sampling (Aghababaei and Makrehchi 2017; see Chap. 9 for technical applications).

According to Wang et al. (2017), conventional crime prediction techniques rely on location-specific historical crime data, which cannot be generalised. However, to forecast crimes directly related to our quality of life, relying on historical crime data alone such as census data has deficiencies, as such data may be limited in scope and fails to capture the full complexity of crimes, which are a function of population and other socio-economic factors (Kadar and Pletikosa 2018). Using aggregate data from multiple sources may offer an opportunity to improve or overcome such deficiencies. According to Kadar and Pletikosa (2018), crafting crime prediction models in accordance with theories of criminology and urban studies requires leveraging large human mobility data (Kadar and Pletikosa 2018). In our research we have taken this into account using from multiple datasets. Hence, being able to forecast crimes in real-time is a significant scientific and sociological problem. To model and predict crimes using historical data, Mohler et al. (2010) used an *Epidemic Type Aftershock Sequence* (ETAS) and Short et al. (2008) developed novel differential equation models for the modelling of crime hotspots dynamics (Wang et al. 2017).

This chapter proposes advancing community policing by creating an integrated hybrid smart crime reporting app that utilises data mined from Twitter streams using sentiment analysis in conjunction with historical police crime datasets. Combining data from the three different sources and integrating them into a Hybrid Smart Crime Reporting App (HIVICRA) creates infographic visual crime hotspot clusters and heat maps that the police could use to forecast crimes.

How Technology Can Improve Community Policing

Both Gladwell (2015) and Mohler et al. (2010) propose that crime is contagious and that its spread leads to the formation of crime clusters in the local environment. Evidence indicates that due to local vulnerabilities known to offenders, criminals may repeatedly attack clusters of nearby targets (Bernasco 2006). Also, a number of studies such as Farrell and Pease (2001), Johnson et al. (2007) and Short et al. (2009) hold the view that crime is contagious and that elevated risks exist for both targets that have been previously burgled and neighbouring houses. Hence, Mohler et al. (2010) suggest using highly clustered event sequences to observe certain types of crime data such as burglary and gang violence, due to their crime specific patterns.

While it is possible to model crime prediction based on either Twitter or historical police crime data separately, relying on one approach alone has its deficiencies (Kadar and Pletikosa 2018). Besides, there are stand-alone crime reporting apps used in numerous parts of the world such as *Citizen app* or *RedZone*, which is a GPS-driven, real-time crime and navigation map app that uses geofencing technology (Prnewswire.com 2018). To take crime prediction models to a higher level, we take a different approach that relies on data extracted from three different datasets and that combines them into a comprehensive infographic visual crime app. Hence, our approach is a holistic way to solve community problems through citizens' involvement with the help of technology, while at the same time encouraging community and police interactions. We thus propose a four-stage hybrid smart app model that draws its data from three different sources. The steps we followed in this process can be summarised as follows:

1. Firstly, we collected neighbourhood crime data from members of the public through the HIVICRA app.
2. Secondly, we collected a sample of Twitter stream data for sentiment analysis using the Tweepy.py utility for 10 h, saving the resulting data in JSON format. The raw data was pre-processed to remove unnecessary data, cleansed, saved and uploaded to a cloud-hosted database. The data cleansing included regularisation and removing redundant data from the tweets to enhance its forecasting abilities.
3. Thirdly, we obtained fiscal historical police crime data about England and Wales from the UK police crime database. The crime data covered the period from March 2016 to March 2017.

4. Fourthly, we incorporated all data into the HIVICRA app to create comprehensive infographic visual crime heat maps and cluster images that police can use for hotspot policing and crime forecasting.

The Value of Twitter Sentiment Analysis

Twitter and big data analytics can be used to extract real-time Twitter feeds using data mining and creating salient visual correlations in the data to help the police predict crime hot spots. Solving problems—be it for business or crime intelligence—requires real-time data solutions. Twitter provides a constant stream of real-time data. The reason is that social media forums such as Twitter are good indicators of what people are thinking about (cp. Chap. 6). Utilising these sources and monitoring them for community policing is a good way to ensure public safety as well as to reach the public. This, however, can be taken to another level by creating an interactive customised predictive policing visual application, using data mined from Twitter feeds in combination with smart apps to create a hybrid solution for crime reporting and forecasting.

The advantage of using Twitter as a source of information is that Twitter provides a wealth of instantaneous up-to-date feeds. Extracting relevant data out of these feeds and applying it to policing may help the police reduce crime. Also, Twitter feeds can alert the police of crimes taking place in real-time, and subsequently police can search for specific hash tags of interest.

This concept of dealing with crimes at their earliest occurrence is related to the successful and influential Broken Windows theory that was used in New York in the 1990s. When New York Police commissioner William Bratton, who trialled this theory, left his job in 1996, New York's murder rate had halved, and misdemeanours had decreased by almost 40%. The rate of felonies had fallen by 44% between 1993 and 1997, and homicide rates declined by 60% (Greene 1999). This was attributed to the effectiveness of the Broken Windows approach (Newyork.cbslocal.com 2017). Therefore, as Judith Greene noted in her book "Zero Tolerance: A Case Study of Police Policies and Practices in New York City", Mayor Rudy Giuliani has credited the reduction in NY crime rates to a "zero tolerance" policing policy approach (Greene 1999). Therefore, adapting such an approach and combining it with the ability to forecast crimes will allow police to identify optimal police beat patrol strategies and to allocate scarce police resources adequately. However, according to Chen et al. (2015), due to lack of criminal predictive factors in the statistical models, conventional crime predicting techniques have some drawbacks and hence fall short in capturing crime incidents (e.g. by relying on only one source of data such as census data or historical police crime data). Such limitations can be overcome by creating hybrid smart crime reporting apps that combine Twitter sentiment analysis with data collected from crimes reported by members of the public as proposed in this chapter.

A hybrid app like ours can play a major role in the implementation of advanced community policing system. However, problems cannot be solved with technology alone. Effective community policing requires management efforts as well as real-time data, as without it police would not be able to achieve the desired results. That is why the majority of UK forces, including the London Metropolitan Police (MET), have a presence on social media and use it on a daily basis to interact with the public. The MET Police has one of UK's most popular police Twitter feeds (Wakefield 2013). However, according to Tim Anderson of the Guardian, many forces are still using outdated equipment (Anderson 2013). He proposed that it is time for police forces to update their technologies, as they seem to lag behind in today's "*always-on-always-connected*" information age, which can give criminals an edge. Embracing new technologies such as smart policing applications and social media gives police officers access to a vast amount of data that is police-relevant, providing police officers with better information and allowing better use of police time and resources (Reyes 2013). Utilising crime intelligence extracted from social networking sites and smart policing apps can help police fight crime, improve community and police relationships, while at the same time allowing police to be more engaged with the public they serve. If citizens are at ease communicating with the police and play their role in policing efforts within their communities by reporting crimes happening in their cities, it would be a win-win situation for both the police and the public.

Furthermore, such an interaction humanises the police and makes them more approachable, since the public can reach them through social media sites such as Twitter or online crime reporting apps. Additionally, this approach creates further "eyes and ears" for the police, when citizens can report crimes through their handheld devices. After all—according to the Broken Windows theory—looking at community problems holistically by paying attention to seemingly small things such as trash and graffiti in neighbourhoods and by dealing with these problems at an early stage is better than being reactive and inviting bigger problems later; because things that may seem small and insignificant at an early stage may add up and lead to bigger problems, which can paint a picture of a neighbourhood that is friendly to crime (Gladwell 2015).

Besides, in times of crisis, social media may well be the best, if not the only possible way for police to reach the public, as a tweet may go a long way or be more effective than messages sent through traditional means. One of the very few empirically based studies available conducted by Xu and Wu (2015) supports that Twitter is a good tool for crisis management. Thus, all organisations including the police that want to function efficiently should embrace the use of social media to connect and engage fully with the communities they protect and serve. Moreover, a lack of police presence in social media may lead others such as criminals to fill their place and spread misinformation, because for criminals or hoaxers operating online traditional hierarchical structures are not necessary or may not exist. Thus, spreading false information is easy, but finding culprits and assigning blame with any certainty would be harder for the police than ever.

According to the 2015 Social Media Survey of the International Association of Chiefs of Police in the USA (IACP 2015), police forces that have used social media have benefitted: In some capacity, 96.4% of police departments in the USA use social media, and more than 85% said that it helped them solve crimes. The survey has also found that for 83.5% of agencies, social media improved police and community relations in their jurisdiction (IACP 2015). In England and Wales, there are 43 disparate constabularies maintaining separate IT system with seemingly little central management apart from DNA databases (Anderson 2013). As Babuta (2017) suggests, British police forces have been using a mix of records from the Police National Computer (PNC), a separate Police National Database (PND), a biometric identification database and the Europe-wide Schengen Information System. According to Babuta, this "*highly localised structure*" seems to have let some police forces to race ahead to embrace new technologies, while "*others lag far behind the curve*" (Babuta 2017). The smart policing HIVICRA app we propose may help bridge this gap. It is a customizable crime reporting smart app that can be used nationwide and in any city around the world.

Description of the Hybrid Infographic Visual Intelligent Crime Reporting Analysis App

In this section, we present the hybrid smart crime reporting app (HIVICRA) that utilises highly clustered event sequences of crime-specific patterns of criminal behaviours that would allow the police to forecast certain types of crimes.

This app emphasises the reduction of crime through citizens' involvement by putting them in charge of reporting crimes occurring in their communities; after all, policing is by consent. Using the HIVICRA app, members of the public can report crimes discreetly from their handheld devices by taking a picture of the crime location and adding a description of the crime, while the app automatically geotags the crime location. HIVICRA also has an in-app-navigation that the police can use to get to the location of crime simply by selecting the option from the crime pointer marker. In the event that the user reporting a crime has no Internet connection, the app also has an offline database storage, which it synchronises automatically when users regain an Internet connection. Crimes reported through the app by members of the public can be delivered directly to the police dispatchers to take appropriate actions.

Our approach takes into consideration that one of the core and basic foundations of community policing is to involve citizens in their own policing. According to Robert R. Friedmann, crime is produced by societal factors over which police has relatively little control (Friedmann 1992). In alignment with this view we created a crime reporting app that focuses on societal problems, which cause crime. This approach puts the public in charge of reporting crimes in their communities, which affect their quality of life and need the attention of the police authorities.

To report crimes, community members have to select the *report crime* button from the app. The user will then be presented with a disclaimer under Criminal Law Act 1967 (CLA 1967). Upon agreeing to the terms of the disclaimer, the user is then presented with options to report crimes such as acid attack, motorcycle theft, and sexual assault. After selecting an option, the user is asked to write a description of the crime witnessed and to take a picture of the crime scene and upload it to the online cloud-hosted database.

The app aggregates data from three different sources (user reports, social media and police database) to create salient visual correlations that

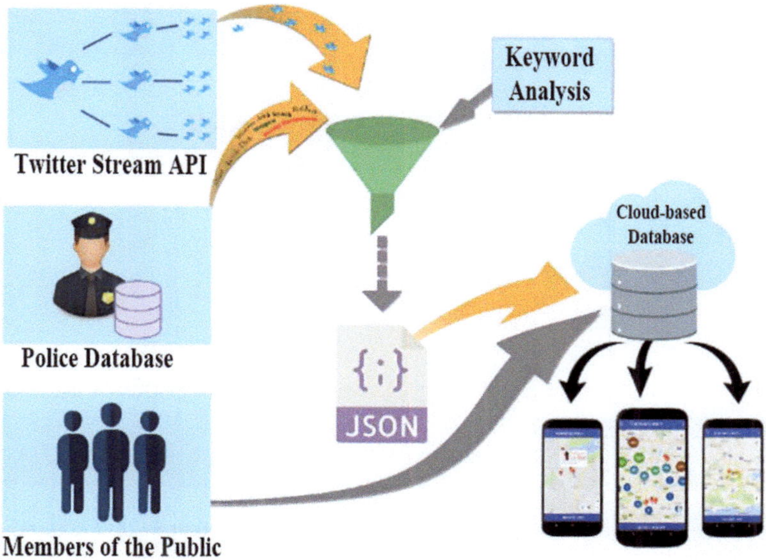

Fig. 10.1 Illustration of the Dataflow in the HIVICRA system showing how data is acquired from various external sources and then filtered before it is transformed into the mobile application

the police can use to forecast crime and for hot spot policing and crime prevention (see Fig. 10.1).

When the HIVICRA app starts, the user is presented with a map showing clusters of crime that were either reported by members of the public or collected through police historical crime datasets or Twitter as shown in Fig. 10.2. By zooming in to the map, more grouped clusters of crime and markers, which represent crime that took place in these localities, are shown. By touching any of the markers, the user will see an infographic visual image of crimes that occurred in this area along with the time, date, type of crimes and any police actions taken against offenders. Also, after touching the marker one of the options presented is an in-app-navigation feature intended for police use to plot the most direct route to the crime location by utilising Google Map app (see Fig. 10.2).

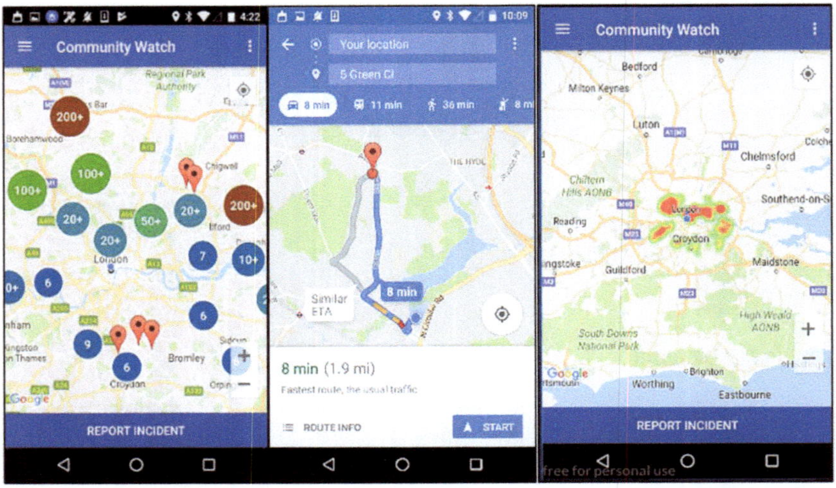

Fig. 10.2 Screenshots of HIVICRA features. From left to right: Clusters of Crime Data Reported, In-App Navigation and Infographic Crime Heat Map

App Features and Functionalities

The following is a complete list of the app's current features and functionalities:

- **Awareness**: The app alerts users to be cautious by vibrating the smart device when they are within a 100-m radius of a crime location. To be vigilant, the app suggests to the user to take necessary precautions.
- **Missing children:** The app can display information about missing children, exploited and trafficked persons together with information about when and where they were last seen.
- **Wanted criminals:** The app can display information about fugitives and persons of police interest such as wanted criminals.
- **In-App Navigation:** This feature presents suggested routes to a crime location.
- **Mitigation Guidance:** The app contains a section that advices users about actions to take in the event of crimes as well as preventive steps when users are in an area known to be crime ridden such as motorcycle gang hotspots.

- **False Reporting:** The app has the ability to ascertain false reports by collecting device information such as phone number, IME number, IP address, email address, geolocation and a device's serial number and make. This information is collected and stored in a database, which will be linked to a specific crime reported. In the event that an individual makes a false report, the police will have the ability to trace it back to the individual who reported it.
- **Visual Cluster and Crime Heat Map:** The heat map we created can be used for forecasting future crimes. Every time, a crime or incident is reported, it is logged and displayed in the heat map. This information is stored online, so that police officers can use it when needed. To avoid that markers overflow the map, we used clusters to group them together.
- **Support for Underground Crime Reporting:** The app has the capability for the public to report in tube stations, where normally there is little cellular as well as GPS signal coverage.

Comparing HIVICRA with Other Mobile Crime Reporting Applications

A crime reporting app with limited functionalities that has some similarities with HIVICRA is an app formerly called *Vigilant*, which was released in October 2016 and has now been rebranded as the *Citizen* app (Perez 2017). Unlike our app, which can be used in any city around the world, the *Citizen* app is region-specific and at the time of writing only available in New York. The app notifies its users when they are near a crime location that is reported to 911. Its features include live-streaming of incidents and user comments. HIVICRA also has the feature to alert its users when they are near a known crime location. Furthermore, unlike the *Citizen* app, which relies mainly on data from 911 calls, HIVICRA utilizes data from members of the public, police crime datasets and Twitter sentiment analysis. Table 10.1 presents a comparison of features of the Citizen and HIVICRA app.

An Example Application of the HIVICRA App

The growing number of social media users and the volume of data shared among them provide predictive modelling opportunities. According to Chen et al. (2015), social media-based predictive modelling is used

Table 10.1 Similarities and differences of two smart applications: HIVICRA and Citizen

App features	*CITIZEN*	*HIVICRA*
In-app navigation	✗	✓
Can be used internationally	✗	✓
Reporting missing children	✗	✓
Reporting most wanted criminals	✗	✓
View crime heat map	✗	✓
Crime clusters	✗	✓
Historical police crime database	✗	✓
Twitter feed crime integration	✗	✓
Metro/underground crime reporting	✗	✓
Live crime/incident broadcasting	✓	✗
911 dispatch integration	✓	✗
Auto geotagging of crime location	✓	✓
False report warning	✓	✓

extensively in fields from elections, stock market trend analysis and product sales to the box office performance of movies to predict the reaction in selected groups. However, studying data collected from selected groups of people alone to predict their future responses may not portray the overall picture of criminal incidents or what is happening in society. Hence, although tweets yield useful insights into issues that are of concern to the public, they cannot be used alone to gauge issues of concern.

Therefore, there is a challenge in relying solely on Twitter sentiments. In addressing these weaknesses and to overcome them, we have used data collected from different sources of information such as police crime databases and combined them with actual crimes reported by the public through the HIVICRA app we created. To find correlations from tweets and to validate them, we focused on new tweets and used keywords or labels for specific crimes we were interested in (e.g. "acid attack"), which were combined with historical crime data collected from the other two sources (i.e. members of the public and the police crime dataset within England and Wales). Taking full advantage of data from these different sources, we generated infographic visual crime analysis clusters and heat maps for crime forecasting and hotspot policing. The primary data source we used for making this crime forecasting model comes from the combination of these three sources. The raw Twitter data collected came from the official Twitter REST API. We have collected this data using a Python-based client script called Tweepy.py.

Database

We obtained our dataset from the England and Wales police crime database from March 2016 to March 2017 in CSV format. The data was then converted to JSON format to be uploaded to our cloud-hosted database, which is accessible to all users of the HIVICRA application as shown in the dataflow system in Fig. 10.1. To collect raw data, we used keyword filtering such as "acid attack", "theft" or "burglary" with a location set to UK. This data was saved in JSON format. We ran the Tweepy.py utility for 10 h to collect 1 GB of data harvesting over 245,000 tweets. We subsequently transferred the data into Excel and cleansed it. Out of the data we collected, only 30% were relevant to our analysis. Therefore, further filtering had to be done to remove excess data leading to the dataset presented in Fig. 10.3.

Analysing the tweets generated in the UK along with their date, time, type of device used to tweet (i.e. Android, iPhone), geolocation (place, co-ordinates) and language used, we found that of the 247,317 tweets we harvested 307 contained incidents of "acid attack" (excluding retweets; see Table 10.2). It is worth noting that the tweet data we collected is reflective of tweets shared within a short period. Nonetheless, it gives insights into the range of public sentiments. Next, we continued our data extraction process through keyword analysis and found that the data contained 7642 incidents of "theft", and 26 incidents of "bicycle theft". "Burglary" incidents amounted to 829 tweets; weapon offences to 9130 tweets, and robberies were tweeted about 3875 times (see Table 10.2).

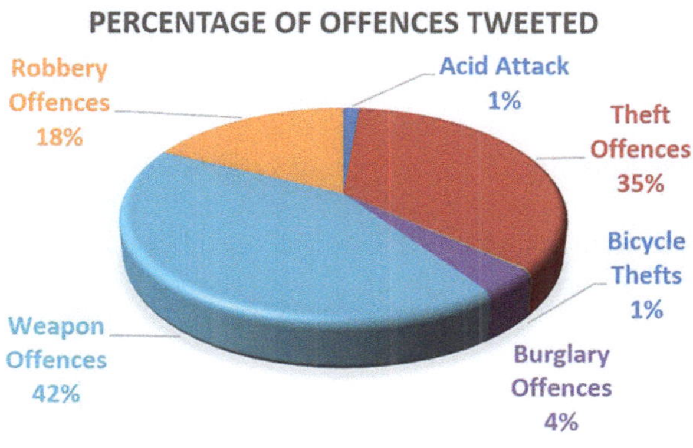

Fig. 10.3 Percentage of the various types of offences reported in Twitter in our dataset

Table 10.2 Categories of offenses and their corresponding number of tweets collected over a 10-h period

Types of offences	*Number of tweets*
Acid attack	307
Bicycle thefts	26
Burglary offences	829
Theft offences	7642
Robbery offences	3875
Weapon offences	9130
Total tweets	***247,317***

Evaluating the Twitter Sentiment Results

After evaluating the results of the incidents tweeted and actual events reported in the mainstream media, we fact-checked a small sample of the tweets against crime events reported in the media on the day we collected the tweet data and found that the Twitter feeds were consistent with the crimes reported by media on that day. Thus, although Twitter feed content alone cannot be relied on to forecast crime, it can serve as a tool for gauging crime events. Furthermore, although the Twitter feed we used was collected within a span of only 10 h, we found that our sentiment analysis can still reveal certain trends. For instance, our Twitter analysis results showed a high number of "acid attack" tweets (307 to be exact).

This is indicative of a trend of corrosive attacks happening in 2017 across London to the point that newspaper outlets were branding London as "the acid capital of Britain" (Robson and Logan 2017). Consequently, the Metropolitan Police faced warnings that acid attacks have the potential to become a "fashionable" crime among Londoners (Crerar 2017). Our Twitter sentiment analysis thus showed that there is a high correlation between the content shared in tweets and actual events occurring in the regions the tweets are coming from.

Discussion and Future Work

Aghababaei and Makrehchi (2017) developed a model for crime trend prediction employing Twitter sampling to identify whether crime rates were increasing or decreasing over a prospective period. Their experiments revealed a correlation between features extracted from Twitter content and the crime rate directions (Aghababaei and Makrehchi 2017).

Conventional crime prediction models rely on historical crime datasets. In this chapter, we presented an innovative hybrid approach to forecast crime in a geographic space using data sources from multiple datasets, in particular anonymised aggregate data of crimes reported by members of the public through the HIVICRA smart crime reporting app we created, in conjunction with data derived from Twitter keyword sentiment analysis and historical police crime data records. The main contribution of this chapter lies in combining anonymised crime data derived from three different datasets to create infographic visual crime clusters and heat maps that the police can use to forecast, prevent and tackle crimes. Using crime clusters and heat maps, for instance, police may predict whether specific area may become crime hotspot or not.

Our hybrid solution is more comprehensive than other related works we have compared it against, as it obtains its data from three different sources rather than relying on Twitter streams alone. It follows similar efforts, such as Chen et al. (2015), who have also used a combination of Twitter sentiment analysis and weather information for crime prediction.

In conclusion, our approach presents a preliminary investigation of how social media sentimental analysis can be used in forecasting crimes. This research shows the applicability of how data mined from social networking sites in conjunction with data from other smart apps and data from historical police crime datasets can be used in community policing. Although, the Twitter feed data we used in this research was small, as it was collected in a span of only 10 hours, and the amount of police crime data we were able to integrate into the HIVICRA app was small too due to database hosting storage limitations, our solution was able to establish correlations and identify promising possible targets for police interventions based on data fed into our solution. Although the relationship between crime and sentiment analysis cannot be directly verified at this stage, our research revealed that combining sentiment analysis with smart crime reporting apps is possible to forecast crime and may have the potential of reducing crime.

Our approach fills a gap by providing a hybrid smart app that has the potential to reduce and improve crime forecasting by using crime intelligence derived from social network sentiment analysis and historical police crime datasets in combination with crimes reported by members of the public. To improve crime forecasting accuracy of the sentiment analysis, we strongly recommend obtaining Twitter stream samples for longer periods. Doing so may improve crime forecasting accuracy.

Overall, integrating crime data from different sources and presenting it into meaningful and easy to understand visual clusters and heat maps can fill a gap in community policing. Here "visual" is not meant in the sense of images but in the form of geotagged map locations with markers classifying crime hotspots and types of crimes occurring in each area. Doing so makes police beat deployment coverage decisions and the monitoring of crimes more efficient, while saving police time and freeing up scarce police resources at a time of austerity measures and police budget cut backs in many cities across Europe.

References

Aghababaei, S., & Makrehchi, M. (2017). *Mining social media content for crime prediction.* 2016 IEEE/WIC/ACM International Conference on Web Intelligence (WI) (pp. 526-531). Available online: https://www.deepdyve.com/lp/institute-of-electrical-and-electronics-engineers/mining-social-media-content-for-crime-prediction-w9xnWWQL20

Anderson, T. (2013). Technology on the beat: How IT can enhance policing. *The Guardian* [online]. Retrieved September 6, 2017, from https://www.theguardian.com/public-leaders-network/2013/dec/18/criminal-justice-technology

Babuta, A. (2017). Big Data and Policing. An Assessment of Law Enforcement Requirements, Expectations and Priorities. *Royal United Services Institutefor Defence and Security Studies* (Occasional Paper). Available online: https://rusi.org/sites/default/files/rusi-bigdata-press-2017.pdf

Bernasco, W. (2006). Co-offending and the choice of target areas in burglary. *Journal of Investigative Psychology and Offender Profiling, 3*(3), 139–155.

Chen, X., Cho, Y., & Jang, S. (2015). *Crime prediction using Twitter sentiment and weather*. IEEE Systems and Information Engineering Design Symposium (pp. 63-68). Available online:https://ieeexplore.ieee.org/stamp/stamp.jsp?tp=&arnumber=7117012

Crerar, P. (2017). *Acid attacks becoming 'fashionable' crime in London, police are warned* [Online]. Retrieved from https://www.standard.co.uk/news/crime/acid-attacks-becoming-fashionable-crime-in-london-police-are-warned-a3592336.html

Farrell, G. & Pease, K. (eds.). (2001). *Repeat Victimization* Monsey, NY: Criminal Justice Press.

Ferreira, B. (1996). *The use and effectiveness of community policing in a democracy* [online]. Retrieved September 6, 2017, from https://www.ncjrs.gov/policing/use139.htm

Friedmann, R. (1992). *Community policing: Comparative perspectives and prospects* (p. 8). New York: St. Martin's Press.

Gladwell, M. (2015). *The tipping point how little things can make a big diffference.* Abacus: New York.

Greene, J. A. (1999). Zero tolerance: A case study of police policies and practices in New York City. *Crime and Delinquency, 45*, 171–182.

Hall, N. (2017). *'Don't visit crime victims, email them,' police chief says* [online]. Retrieved September 12, 2017, from https://www.rt.com/uk/402893-police-victim-email-crime/

IACP. (2015). *International Association of Chiefs of Police 2013 Social Media Survey Results* [online]. Retrieved September 17, 2017, from http://www.iacpsocialmedia.org/wp-content/uploads/2017/01/FULL-2015-Social-Media-Survey-Results.compressed.pdf

Johnson, S., Bernasco, W., Bowers, K., Elffers, H., Ratcliffe, J., Rengert, G., & Townsley, M. (2007). Space–time patterns of risk: A cross national assessment of residential burglary victimization. *Journal of Quantitative Criminology Criminology, 23*(3), 201–219.

Kadar, C., & Pletikosa, I. (2018). *Mining large-scale human mobility data for long-term crime prediction.* EPJ Data Science, 7, Article number 26. Available online: https://epjdatascience.springeropen.com/articles/10.1140/epjds/s13688-018-0150-z.

Meyer, M., & Graan, J. (2011). Effective community policing in practice: The Roodekrans neighbourhood watch case study, West Rand. *South African Journal of Criminology, 24*(2), 1–2.

Mohler, G., Short, M., Brantingham, P., Schoenberg, F., & Tita, G. (2010). Self-exciting point process modeling of crime. *Journal of the American Statistical Association, 106*(493), 100-108.

Newyork.cbslocal.com. (2016). *Ceremonial send-off marks Bratton's Last Day on the Job as Police Commissioner* [online]. Retrieved September 6, 2017, from http://newyork.cbslocal.com/2016/09/16/nypd-bratton-ceremonial-send-off/

Omand, D., Miller, C., & Bartlett, J. (2014). Towards the discipline of social media intelligence. In C. Hobbs, M. Moran, & D. Salisbury (Eds.), *Open Source Intelligence in the Twenty-First Century. New security challenges.* London: Palgrave Macmillan.

Perez, S. (2017). Banned crime reporting app Vigilante returns as Citizen, says its 'report incident' feature will be pulled. *TechCrunch.* Available online: Retrieved September 11, 2017, from https://techcrunch.com/2017/03/10/banned-crime-reporting-app-vigilante-returns-as-citizen-says-its-report-incident-feature-will-be-pulled/

Reyes, J. (2013). *92% of U.S. law enforcement agencies use social media Technical.ly.* Available online: Retrieved August 23, 2017, from https://technical.ly/2013/10/23/police-social-media

Robson, S., & Logan, R. (2017). Five acid attacks by 'moped thieves' latest in epidemic hitting London streets. *Mirror* Available online: https://www.mirror.co.uk/news/uk-news/london-acid-attack-capital-uk-10795265

RT International. (2017). *Police* threaten to arrest minors and evict their parents in crime crackdown. RT (Russia Today). Available online: https://www.rt.com/uk/402355-police-manchester-arrest-parents/

Short, M. B., D'Orsogna, M. R., Brantingham, P. J., & Tita, G. E. (2009). Measuring and modeling repeat and near-repeat burglary effects. *Journal of Quantitative Criminology, 25*(3), 325–339.

Short, M., D'Orsogna, M., Pasour, V., Tita, G., Brantingham, P., Bertozzi, A., & Chayes, L. (2008). A statistical model of criminal behavior. *Mathematical Models and Methods in Applied Sciences,18*(1), 1249–1267.

Telep, C., & Hibdon, J. (2018). Community crime prevention in high-crime areas: The Seattle Neighborhood Group Hot Spots Project. *City & Community, 17*(4), 1143–1167. https://doi.org/10.1111/cico.12342

University of Portsmouth. (2013). *Models of policing* Available online: http://compass.port.ac.uk/UoP/file/7588425b-136c-40fc-a657-1dc4a13cb05c/1/Models%20of%20Policing_IMSLRN.zip/page_03.htm

Wakefield, J. (2013). Future cops: How technology is set to change policing. *BBC news/technology* Available online: https://www.bbc.co.uk/news/technology-22954783

Wang, B, Zhang, D., Zhang, D., Brantingham, P. J., & Bertozzi, A. L. (2017). *Deep learning for real time crime forecasting* Available online: https://arxiv.org/pdf/1707.03340.pdf

Xu, J., & Wu, Y. (2015). Using Twitter in crisis management for organizations bearing different country-of-origin perceptions. *Journal of Communication Management, 19*(3), 239–253.

CHAPTER 11

Contextual Visualization of Crime Matching Through Interactive Clustering and Bayesian Theory

Nadeem Qazi and B. L. *William Wong*

Introduction

Police and law enforcement agencies these days also perform social media analysis to gain a better understanding of criminal social network structures and to identify potential criminal activities and criminal networks. The use of data mining techniques in social media analysis, however, faces issues and challenges such as linkage-based structural analysis, association extraction, community or group detection, behavior and mood analysis, sentiment analysis and dynamic analysis of streaming networks. In this chapter, we focus on two main issues, that is, association and community detection particularly focusing on the crime matching process.

N. Qazi (✉)
Department of Mechanical, Aerospace, and Civil Engineering, College of Engineering, Design and Physical Sciences, Brunel University, London, UK
e-mail: Nadeem.qazi@brunel.ac.uk

B. L. W. Wong
Middlesex University, London, UK
e-mail: w.wong@mdx.ac.uk

B. Akhgar et al. (eds.), *Social Media Strategy in Policing*,
Security Informatics and Law Enforcement,
https://doi.org/10.1007/978-3-030-22002-0_11

Analysts during crime matching assign criminals to previously solved or unsolved crimes, establishing an association between them. They usually employ crime variables such as time, location, modus operandi of the committed crime and the profile of the offender to answer a question such as "Given a Y crime identified by a modus operandi pattern, who could possibly be X offender?" Alternatively, in another approach of crime matching, they possibly try to answer the question "For given an X offender, what could be possible Y unsolved crimes, that could be potentially thought to be committed by him based on similarity patterns of the unsolved crime patterns and the suspects?" They possibly try to find out what are the chances that an offender is responsible for an unsolved offense.

This process of crime matching is indeed an information-intensive querying process that requires establishing associations among crime entities to discover and reconstruct crimes through the analysis of the evidence left at the crime scene. Analysts spend a large amount of time performing extensive database searches, reading crime reports looking for clues for associations among entities such as criminals, vehicles, weapons, bank accounts and organizations. The resultant data from these searches often requires a tedious job of grouping the crimes based on crime pattern similarities. In addition, this process also lacks appropriate visualization to help the cognitive thinking process. It therefore causes analysts to face a number of significant difficulties including making sense of collated data, distinguishing the relevance or similarities among the cases and identifying and understanding associations between criminal entities.

In our earlier research towards the criminal analysis (Wong and Kodagoda 2016), we observed that police analysts follow a search based on associative questionings for establishing associations among the criminal entities to discover and reconstruct crime scenes. For example, analysts studying a given crime pattern may be interested in generating contextual, cognitive and domain-specific queries such as "Who else has used the same modus operandi in the similar crime or Who are the other offenders who have committed similar crimes?" This search, however, is different than traditional keyword or semantic based searches and requires special search mechanisms.

Following this need in our earlier research, we introduced the *associative search mechanism* in digital criminal analysis (Qazi et al. 2016). An associative search, unlike keyword and semantic search, searches along the networks of associations between objects such as people, places, other organizations, products, events, services and so forth. The critical issue in an associative search is to establish the scope or level of associations among related concepts over the given dataset. We proposed a 5WH-model comprising five general but associated concepts, each representing a question, which are linked to each other through a set of properties or attributes to reveal the relationship, association or relevancy among crime concepts. These associated concepts are WHAT (what has happened, i.e. type of offense), WHO (who has committed the crime identifying offender or group of offenders), WHEN (when an offense happened), WHERE (the geospatial information about the offense) and HOW (the modus operandi used in committing the crime). We employed this associative search mechanism to develop a visualization tool that assists in discovering a list of possible suspects for an unsolved crime.

In this chapter, we describe the extension of our framework and propose an association model for extracting multilevel associations based on associative questioning. We also describe data mining techniques used to visualize these associations through a 2D crime cluster space. Each of these clusters groups the solved and unsolved crimes with associated perpetrators based on spatial, temporal and modus operandi similarity of the crimes. Additionally, we also attempt to answer the following question for the crime matching process: "Is it possible for a suspect to have committed a particular unsolved crime?" We answer this question by identifying and extracting all possible linkages between a suspect and unsolved crimes based on spatial and modus operandi characteristics. We further demonstrate the use of the Naive Bayes theorem to determine the likelihood of an offender being linked with unsolved crimes. The developed framework provides a complete data analytic solution towards identifying and understanding associations between crime entities and thus expedites the crime matching process.

The rest of the chapter is organized as follows: "Related Work "section presents related research followed by a dataset description in section "Dataset". We discuss the proposed model for extracting the multilevel associations for crime matching in section "Multilevel Association Model". Section "Plausible Unsolved Crime List" describes the process of associating a suspect with unsolved crimes through Bayesian theory. "Association Visualization" section presents the association visualization technique through clustering and treemaps. Conclusions are drawn in the final section of the chapter.

Related Work

The work presented in this chapter integrates multiple domains in data science for criminal analysis under a single framework. It includes associative search or link analysis for finding a relationship between crime entities, the Naive Bayes algorithm for determining the likelihood of criminal associations with unsolved crimes, clustering for identifying similar crimes and lastly network or tree visualization.

Researchers have demonstrated the use of these domains separately to solve different criminal analysis problems. Qazi et al. (2016), for instance, introduced a formal concept-based model to reveal the associations highlighting hotspots, offender profiles and associated offenders in a criminal activity. Other researchers such as Oh and Cho (2010) have developed a semantic network of mobile logs based on associative search. The network offers interactive visualizations connecting data such as GPS, Call, SMS, picture viewer, MP3, charging and photo tagging. In the software engineering domain, Takuya and Masuhara (2011) have developed a source code recommendation tool based on an associative search engine called *GEETA*. It spontaneously searches and displays example programs while the developer is editing a program text.

In criminal investigation domain, Thongsatapornwatana and Chuenmanus (2014) have employed association analysis with a reputation algorithm to identify vehicles that are potentially involved in criminal activities. They used the vehicle and criminal activity dataset of the Defense Technology Institute (DTI), Thailand. Their approach employed the matching of license plates, color, brand and type of target vehicle with that of a 1-year checkpoint crossing dataset to establish the criminal activity of target vehicles. Another interesting system is *JIGSAW* by Stasko et al. (2008) that extracts relevant entities such as people, addresses and dates from a collection of documents and shows connections between these entities. It uses the node-link diagram in a basic graph view to display connections between entities and documents, thus allowing analysts to explore the documents.

Recently, criminal network analysis has received great attention from researchers. Rasheed and Wiil (2015) have proposed a framework called *PEVNET* to analyse intra-network criminal activities. The framework offers visualization of collaborations amongst individual criminal nodes and detects the patterns of hidden criminal activities in the sub-clusters along with trend analyses. Dynalink from Park et al. (2012) is a framework for visualizing criminal networks using an animation approach to visualize the changes of networks over time, thus assisting in discovering and analysing both relational and temporal patterns of criminal networks. Isah et al. (2015) have used bipartite network models for extracting hidden ties in both traditional and cybercrimes for pharmaceutical crimes and underground forum datasets, respectively. Among other systems that use network visualization tools, we mention *Netmap*, *Analysts Notebook* and

Coplink by Hauck et al. (2002), research by Ding et al. (2009), *RECAP* by Brown (1998) and *CrimeLink Explorer* by Schroeder et al. (2007).

Clustering techniques either independently or in combination with other data mining techniques such as self-organizing maps (Alruily et al. 2010) have been utilized for some time in criminal investigations for grouping similar crimes. For example, Nath (2006) has utilized a k-means clustering algorithm on real-time data acquired from police forces to find crime patterns, while Le-Khac et al. (2016) have employed clustering techniques to detect money laundering in the banking industry. Didimo et al. (2011) employed k-cores hierarchical clustering to visualize financial activity networks connecting entities like bank accounts, addresses, amount and types of transactions, and motivations extracted from financial reports. In another approach, Borg et al. (2014) have detected series of residential burglaries through minimum cut based graph clustering. They used a feature vector consisting of modus operandi, residential characteristics, stolen goods and spatial similarity to group similar crimes. Reich and Porter (2015) proposed a semi-supervised clustering approach based on the Bayesian model utilizing crime locations and offenders modus operandi for burglary crime series identification. Keyvanpour et al. (2011) performed crime data clustering for crime matching employing neural networks to mine attribute maps and a k-means algorithm to group the output. Bsoul et al. (2013) detected crime patterns on the news through document clustering using 10 types of crime including traffic violations, thefts, sex crimes, murders, kidnappings, fraud, drugs, cybercrime and arson. They employed an affinity propagation algorithm for determining the number of clusters to be fed into the k-means algorithm.

There is also a growing trend in criminal analysis towards the use of Naïve Bayes to model crime linkage. Zoete et al. (2015) proposed a crime linkage model based on Bayesian networks to show how evidence observed in one crime can be used in another and vice versa. Their proposed crime linkage model, however, simplifies the reality and hence does not capture all the problems that play a role when linking crimes. Working in the same direction, de Zoete et al. (2017) employed Bayesian networks for modeling multiple offenders for two separate offenses. They discussed a mock case example to show that subtle differences between situations can lead to substantially different conclusions in terms of posterior probabilities of a certain suspect being one of the offenders in a particular crime. Vural and G¨ok (2017) constructed a naive Bayesian model over synthetically generated incident-level crime data to express the probability of criminals for committing a crime. They used crime date, location and the criminals' names and acquaintances as clues to predict the posterior probability of a criminal to be associated with an unsolved crime.

The work cited above, however, either lacks appropriate visualization for associations or lacks knowledge discovery abilities for association identification and requires this information to be provided by the user to allow visualization. In this chapter, we combine interactive dynamic visualization along with knowledge discovery abilities of multilevel criminal associations into a single envelope over an anonymized burglary data set.

Dataset

We used the Anonymized Burglary dataset consisting of 164,800 recorded cases of five different types of burglaries, collected from UK law enforcement agencies to evaluate the performance of our proposed associative search mechanism. A crime report document is generally defined to be a logical unit of textual data consisting of the crime reference, offender information, modus operandi description, offense category, location, time and date of the crime along with related information. ***Modus operandi*** is the method of the operation adopted to commit a crime, which can refer to preparation actions, crime methods, the weapon used the position of entry, etc. Twelve modus operandi variables including entry position, entry type, fixture, fixture type, location, search type, exit type, and exit fixture were used for each of the criminal cases. Each of these modus operandi variables has a set of predefined values for selection in crime report filing.

Multilevel Association Model

The association mining task includes object ranking, group detection, collective classification, link prediction and subgraph discovery. During the investigation the explicit social links such as friendship, financial exchange and telephonic connection are used to detect a group or network. However, the implicit crime links derived from modus operandi, geographic relationship, crime and time of the day are also very valuable towards solving a crime. Our proposed model elicits such implicit hidden relationships from the crime report and is particularly suited for the group detection tasks of association mining. The model shown in Fig. 11.1 elicits multilevel associations from crime texts based on the 5-WH questions presented in Table 11.1. These associations elicited through rule-based heuristics

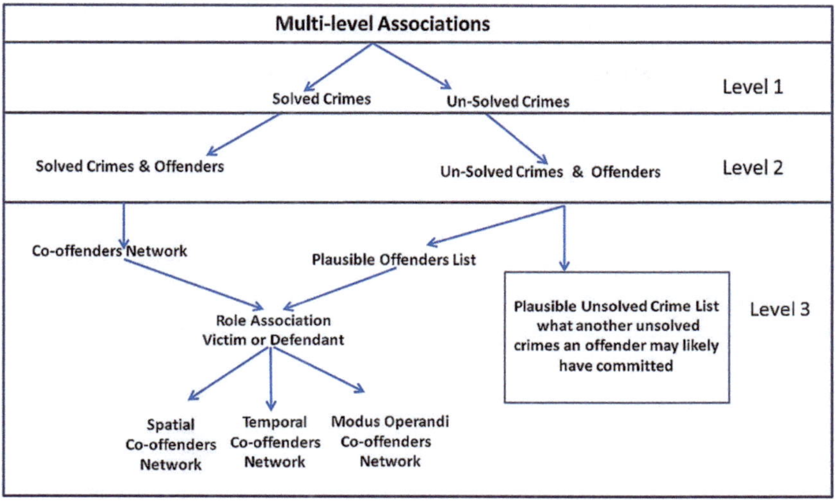

Fig. 11.1 Multilevel associations model

Table 11.1 5 WH-based associative questions

1. What other offenses have occurred for a given crime pattern?
2. Where else have crimes like these been committed?
3. Who are the known offenders operating in an area, and what is their modus operandi?
4. Who else in the past has committed crimes like this?
5. How often have offenses like the given crime pattern occurred?
6. What are the other modi operandi that have been used in committing crimes of this type?
7. What are the additional details of the associated offenders/victims, his/her past history, etc.?
8. What are the geospatial profiles of offenders, including their temporal, spatial and other similar criminal activities resembling the given crime pattern?
9. How many times has the offender committed similar crimes, and what are their temporal and spatial details?
10. What is his/her pattern of modus operandi?
11. Where is an offender most likely to commit an offense, and who else has committed the same crime at this location?

and similarity matching reveal crime associations and group identifications in the linked crime dataset.

The crime associations identify solved and unsolved crimes and the associated offenders or victims as depicted in level 1 and level 2 of the association hierarchy in Fig. 11.1. The third level of associations in our model reveals communities based on spatial–temporal and modus operandi-based similarities including a co-offender network, a plausible suspect list for unsolved crimes and a plausible unsolved crimes list for a given offender. Each of these networks or communities is created from heterogeneous types of nodes and edges. These nodes may be perpetrator, location, offense, time or modus operandi. The edges of the network connect these nodes on the basis of similarities in spatial, temporal and modus operandi characteristics. Following the notion of heterogeneous representations of linkages (Sun 2013), we named these networks *spatiotemporal modus operandi (STM) network* and employed acyclic graphs to represent them.

The root and leaf nodes of each of these STM networks are different. For example, the STM network graph for the plausible suspect list takes a given unsolved crime as the root node and extracts names of the suspects for unsolved crimes as leaf nodes of the network. Likewise, for the co-offender network the root node is the offender and the leaf nodes are its associated co-offender nodes of the network. Finally, for the plausible unsolved crimes list, which is the focus of this chapter, the root node is the offender and leaf nodes would be unsolved crimes that may be committed by the offender.

Plausible Unsolved Crime List

In our earlier study (Qazi and Wong 2017), we described a plausible suspect list to answer the first question of the crime matching process: "Who might have committed the given unsolved crime?" In this chapter, we focus on the plausible unsolved crimes list to answer the second questions of crime matching: "Is it possible for a suspect to have committed a particular unsolved crime?"

We define this problem as finding the most likely unsolved crime(s) for a given criminal when the history of the criminal is known given incident-level crime data. The basic idea is to look into the likelihood of unsolved crimes to be committed by an offender based on the similarity of crime patterns in unsolved crimes. This likelihood depends on the degree of similarity/distinctiveness and can be measured using the Bayes theorem.

Table 11.2 Associative questions for crime matching

1. What are the crimes that have occurred in the criminal area of the suspect?
2. What is the similarity of unsolved crimes reported in that area to that of the crime pattern of the suspect?
3. Who are the other offenders who have committed crimes in the vicinity of the area and time, and what is their similarity with that of the unsolved crime(s)?

In a firsts step, we elicit the plausible crime list that may be committed by a suspect employing similarity matching of the crime components of the suspect with that of unsolved crimes through associative questioning (see Table 11.2). We then measure the likelihood of the suspect committing the unsolved crime(s) through Bayesian theory.

Bayes theorem describes the conditional probability of an event's occurrence related to any condition; in other words, it gives the probability that something will happen, given that something else has already occurred. It follows from the axioms of conditional probability but can be used to powerfully reason about a wide range of problems involving belief updates. Given a hypothesis and evidence, the Bayes theorem provides a way of calculating a posterior probability given by the following equation:

$$p(H \mid E) = \frac{p(E \mid H) * p(H)}{p(E)}$$

where

H is the hypothesis or initial belief, which could be either true or false (e.g. the defendant is innocent),
$p(H)$ is the probability of an initial belief, also called prior probability of H,
$p(E|H)$ is called likelihood, and it answers the question: "What is the probability of seeing the evidence given our assumed hypothesis H?"
$p(E)$ is the probability of the evidence irrespective of our knowledge about H,
$p(H|E)$ is the revised or posterior probability of H, given the evidence.

We calculated the posterior probability of all offenders that have shown a similarity to the unsolved crimes in the dataset. We considered the spatial, temporal and modus operandi used in committing a crime as evidence and set the hypothesis that a given suspect has committed the unsolved crime. The evidence vector in our case consists of point of entry, method of entry, location of search, type of search, point of exit, fixture, fixture type, fixture material, day of the week and time of day as given in a crime report. We, however, made an assumption that all of the evidence is conditionally independent of each other to use the Bayesian rule for probability.

All of the evidence features in our case are categorical in nature. Therefore, we first created a frequency table for all evidence from the solved crime data and then calculated the prior probability of the evidence, their likelihood for a given offender for burglary crimes along with the prior probability of the offender committing the burglary. This allows us to answer following questions:

- What is the probability that a suspect may commit a burglary?
- What is the probability that a suspect has committed a burglary in the same spatial location where an unsolved crime has occurred?
- What is the probability of using a specific modus operandi by a suspect in committing a burglary?
- How many similar crimes has he/she already committed?

Finally, following the above-mentioned method, we calculated posterior probabilities of all suspects who show a similarity in their crime pattern to an unsolved crime. Examples are presented in Table 11.3. A posterior probability calculated according this model can be helpful in uncovering the interesting aspects of the reasoning for linking unsolved crimes with a known offender. However, we acknowledge that it does not capture all the problems that play a role when associating a crime with an offender, as the current model depends on the given dataset and lacks relevant information that plays a role when linking crimes to offenders. Still, we think it may be helpful to uncover the interesting aspects of the reasoning for crime matching.

Table 11.3 Association probabilities of suspects with unsolved crimes

Offenders	*Unsolved Crimes*		
	ID:191657520	*ID:181908600*	*ID:251771339*
Offender 1	0.698	0.217	0.901
Offender 2	0.757	0.657	0.815
Offender 3	0.857	0.709	0.215

Association Visualization

Crimes have multidimensional variables consisting of key process indicators (KPIs) such as total number of crimes, number of solved and unsolved crimes, associated offenders, proximity of the offence (represented through postcodes), modus operandi used in committing an offence such as entry position, fixture type used, search location and entry type. Aiming towards both explanatory and exploratory data criminal analysis, we employed a partition clustering technique and tree visualization to present the elicited associations in conjunction with an overview and detail technique. It is implemented by clustering a 2D reconfigurable crime cluster space together with two separate panels, one for an overview, the other for showing details. The basic idea is to present the overall pattern of the crime data in the overview panel and detailed information of each of the objects in the detailed panel. The construction of the 2D crime space through clustering is described in the next section.

Spatial–Temporal and Behavioural Similarity-Based Crime Clustering

Clustering is an unsupervised machine learning technique that groups data on the basis of similarity. It provides a multidimensional analysis by partitioning the collection of objects into groups. Each of these groups contains objects that are more similar to objects in this group than in other groups. Clustering methods can be classified as partition, hierarchical, density and constraint-based methods. In this work we used a partitioning clustering approach. The *partition-based clustering method* classifies data into k non-overlapping groups such that each object belongs to only one group and each group must have at least one object inside in it. K-mean and k-mediods are the most commonly used partition methods for clustering.

We selected spatial, temporal and modus operandi components of the crime patterns as an investigative lens; representing *location* through postcode, street and town, *temporal information* through month, day and time of the offense and *modus operandi* by a set of the predefined variables mentioned above. Unlike the prevalent trend in crime clustering, the idea of dynamic feature selection was employed for interactive clustering. This enables a user to redefine the feature vector selecting or de-selecting any temporal, spatial or modus operandi variables to re-cluster crimes. We performed hot encoding of all categorical data into their numerical representation before feeding the data to a k-mean algorithm.

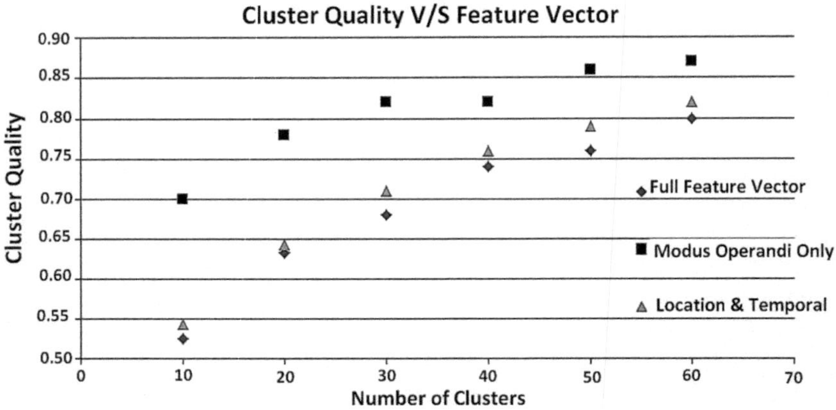

Fig. 11.2 Effect of feature vector on cluster quality

The k-mean clustering generated clusters based on the similarity of the chosen feature vector, grouping solved and unsolved crimes along with associated offenders in each cluster. The quality of each cluster was measured using a similarity score based on the sum of similarity distances of each cluster member to its centroid. We also measured the effect of the attributes in clustering and the cluster quality (see Fig. 11.2). The dynamic configuration of the feature vector helps in determining the effect of crime attributes in clustering. We found that clusters were more homogeneous when only modus operandi was used as a feature vector in clustering.

Additionally, we projected these clusters on a two-dimensional crime space with reconfigurable X and Y-axes; that is, each of the KPIs can be set on either the X or Y-axis, thus enabling the analyst to observe the relationship of KPIs with respect to each other. For example, if a user sets the global similarity on the X-axis and the total number of offenders in a cluster on the Y-axis as shown in Fig. 11.3, then the clusters would arrange themselves revealing how the total numbers of offenders are distributed over cluster spaces as a function of the clusters' global similarity. In addition, a user can also set other KPIs such as proximity, total crime and total offenders on any of this axis to reveal hidden relationships on a two-dimensional crime space. In this way, any KPI can be set on either the X or the Y-axis to visualize crime clusters in user-specified dimensions revealing more insights into the data. For example, when proximity is chosen on the

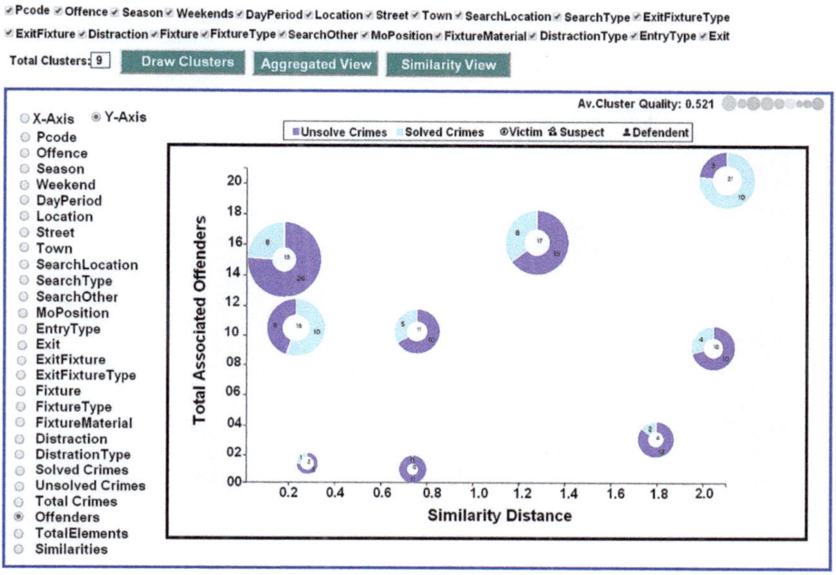

Fig. 11.3 Overview panel of the crime clustered space

X-axis and crime population on the Y-axis, a user can spot the association with between proximity and the criminal population on the 2D-crime space.

This 2D crime cluster space following the "overview and detail" visualization technique has two panels: an overview panel (Fig. 11.3) and a detailed panel (Fig. 11.4). The overview panel represents the overall trend of the solved and unsolved crimes and associated offenders inside every cluster. Each of the clusters is represented visually as doughnut with two arc lengths proportional to the number of solved and unsolved crimes, whereas an icon glyph is used to visualize offenders inside the doughnuts. The radius of the doughnuts is kept proportional to the sum of solved and unsolved crimes and associated offenders so that bigger doughnuts represent a larger population of the crime KPIs.

The detailed panel of the crime space depicts the similarity of crimes with each other. Each cluster in this panel is represented as a big circle, showing solved and unsolved crimes as a turquoise and purple circle, respectively. The offenders associated with solved crimes are shown through icon glyphs. Hovering on each of the solved and unsolved crimes shows the details of each crime in a rectangular box. Figure 11.4 shows a

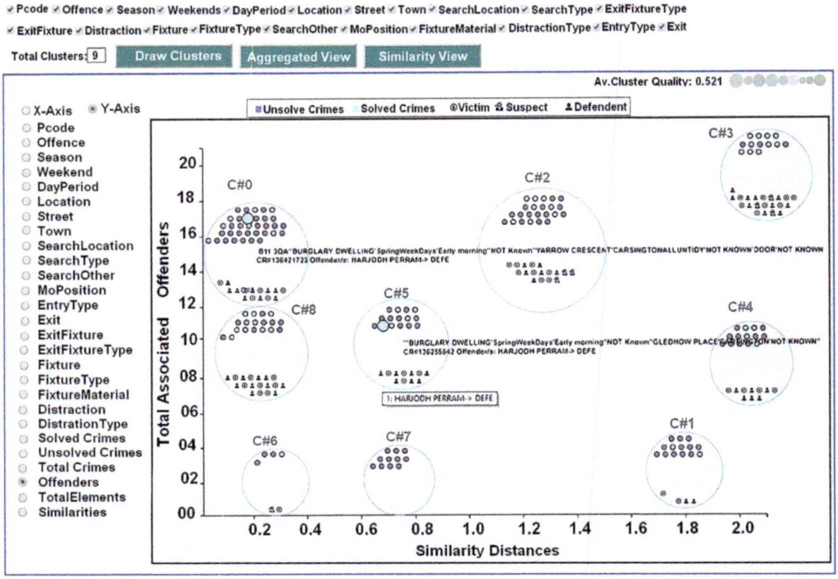

Fig. 11.4 Detailed panel of the crime clustered space

screenshot when an offender is hovered over. Associated solved crimes are highlighted by increasing the size of the solved crime circles in the respective clusters; this goes back to normal when the mouse pointer is removed. This detailed view encourages the analysts to seek further insights into specific crimes and relations between KPIs. Once a solved crime is spotted to match with an unsolved crime, the analyst may want to see the details of the associated offender such as their criminal history and criminal network. For this purpose, we employ knowledge graphs, discussed in the next section.

Knowledge Graphs

We present criminal associations in the form of a heterogeneous network called *knowledge graph*. This graph presents associations in a parent–child relationship creating a heterogeneous radial tree graph structure (see Fig. 11.5). This tree structure clusters the perpetrators into three main groups corresponding to spatial, temporal and modus operandi similarities, showing linkages between a chosen root node and other linked entities.

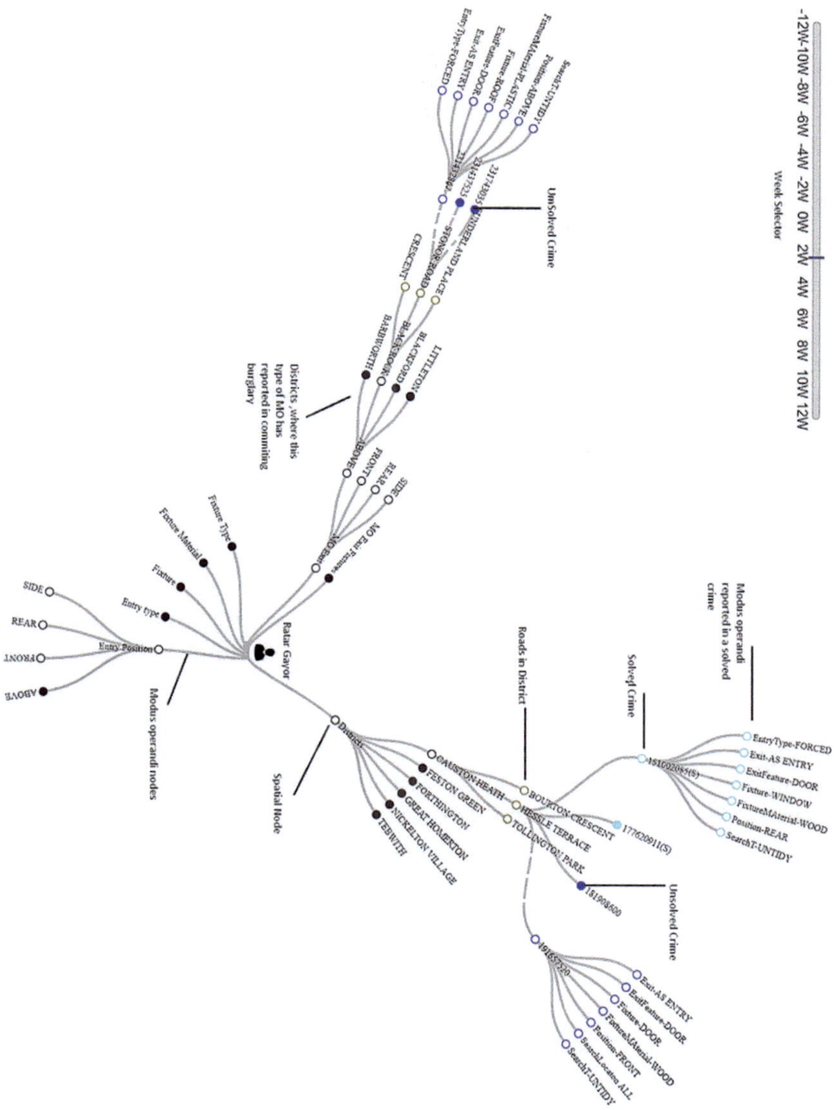

Fig. 11.5 Radial knowledge graph for the plausible unsolved crime list

The root and child nodes are collapsible and are represented by iconic graphics, while thick and broken lines are used to show the type of relationship. These collapsible nodes, when clicked, expand to show the level of similarity. Solid nodes indicate that they have children and can be expanded; expanded nodes are shown hollow, when they have no further children.

The root node for the plausible crime list represents the suspect with a graphic icon. The root node (selected offender) is branched into child nodes, one each for the spatial and modus operandi component. The spatial component is further resolved up to two levels narrowing down the similarity of the spatial information to districts and streets. All unsolved/solved crimes that occurred in these districts and streets are presented using purple and turquoise circles, respectively. The temporal component of crime patterns of given suspects can be manipulated by a slider bar, which is distributed in intervals of 2 weeks. This slider bar can be used to filter the data before and after the date of crime committed by a suspect. The week zero (0 W) refers to the actual date the crime was committed, as reported in the crime report. Whenever a user moves the slider bar, the data is filtered to the chosen interval and the similarity together with the posterior probability are updated for the filtered data. Lastly, the modus operandi node of the knowledge graph compares the behavioural components of the unsolved crime(s) with that of a suspect's pattern, starting from district and street; its subsequent levels narrow the crime pattern down for each element of the modus operandi, showing similar solved/unsolved crimes.

Figure 11.5 shows plausible unsolved crimes for the Offender1. He has committed burglaries in several districts (dark-grey nodes), namely, Causton Heath, Feston green, Forthington, Great Homerton, Nickelton Village and Tebwith. One of the expanded district nodes (showing hollow) present three streets, namely, Bourton Crescent, Hessle Terrace and Tollington Park, where this offender committed a crime. On Hessle Terrace (yellow hollow circle), he has committed a solved crime (id 177620911S, turquoise colored circle). However, there is an unsolved crime reported on the same street, which is shown in purple colour, as based on similarity measures, it may be thought to be committed by the same offender (indicated by broken lines). The modus operandi of the two crimes (solved and plausible unsolved) can be seen and compared for similarity as well. The posterior probability of committing this crime based on modus operandi and the spatial similarity is 0.698 (see Table 11.3). These kinds of associations may support the thinking and reasoning strategies that analysts adopt during an investigation.

Conclusion

In this chapter, we have shown how associative search can aid analysts in crime matching and presented a model for the elicitation of multilevel associations using 5WH questions for a burglary dataset. We have merged temporal–spatial and modus operandi characteristics in one envelope to identify these associations in a form of heterogeneous network that could take associations for a given type of input node, highlighting either an offender network, a plausible suspect list or an unsolved crime list depending on the nature of the input node. We have projected these associations using interactive clustering to visualize networks of associations between criminal entities. We have further evaluated associations of the suspect with unsolved crimes through naive Bayes theory calculating the posterior probability of an offender to be associated with an unsolved crime, using similarity of crime patterns as evidence.

We do acknowledge that our approach does not capture all problems that play a role when associating a crime with an offender. Still, we think our method may be helpful to uncover interesting aspects for reasoning in crime matching. The key to this research is the belief that there exist possible relationships within the various crime dataset, and that simple visualizations of underlying associations can be helpful for the analytical reasoning during a crime matching process. Such associations can provide the basis for activating ideas, thoughts and tentative or plausible conclusions that could trigger new lines of inquiry. Our framework thus enables crime analysts to make assessments rather than a recommendation and to act on evidence only as appropriate. Police analysts during preliminary user feedback sessions have given positive feedback, indicating that this prototype has the potential to improve the efficiency of the criminal investigation process.

Acknowledgements The research leading to the results reported here has received funding from the European Union Seventh Framework Programme through Project VALCRI, European Commission Grant Agreement N FP7-IP-608142, awarded to Middlesex University and partners.

References

Alruily, M., Ayesh, A., & Al-Marghilani, A. (2010). Using self organizing map to cluster Arabic crime documents. In *Computer science and information technology (IMCSIT), proceedings of the 2010 International Multiconference on*. IEEE, pp. 357–363.

Borg, A., Boldt, M., Lavesson, N., Melander, U., & Boeva, V. (2014). Detecting serial residential burglaries using clustering. *Expert Systems with Applications, 41*(11), 5252–5266.
Brown, D. E. (1998). The regional crime analysis program (recap): A framework for mining data to catch criminals. In *Systems, Man, and Cybernetics, 1998 IEEE International Conference on*, vol. 3. IEEE, 1998, pp. 2848–2853.
Bsoul, Q., Salim, J., & Zakaria, L. Q. (2013). An intelligent document clustering approach to detect crime patterns. *Procedia Technology, 11*, 1181–1187, 2013, 4th International Conference on Electrical Engineering and Informatics, ICEEI 2013
de Zoete, J., Sjerps, M., Lagnado, D., & Fenton, N. (2015). Modelling crime linkage with Bayesian networks. *Science & Justice, 55*(3), 209–217.
de Zoete, J., Sjerps, M., & Meester, R. (2017). Evaluating evidence in linked crimes with multiple offenders. *Science Justice, 57*(3), 228–238.
Didimo, W., Liotta, G., Montecchiani, F., & Palladino, P. (2011). An advanced network visualization system for financial crime detection. In *2011 IEEE Pacific visualization symposium*, pp. 203–210.
Ding, L., Steil, D., Hudnall, M., Dixon, B., Smith, R., Brown, D., & Parrish, A. (2009). Perpsearch: An integrated crime detection system. In *Proceedings of the 2009 IEEE international conference on intelligence and security informatics, ser. ISI'09*. Piscataway, NJ: IEEE Press, 2009, pp. 161–163
Hauck, R. V., Atabakhsh, H., Ongvasith, P., Gupta, H., & Chen, H. (2002). Using coplink to analyze criminal-justice data. *Computer, 35*(3), 30–37.
Isah, H., Neagu, D., & Trundle, P. (2015). Bipartite network model for inferring hidden ties in crime data. In *2015 IEEE/ACM international conference on advances in social networks analysis and mining (ASONAM)*, pp. 994–1001.
Keyvanpour, M. R., Javideh, M., & Ebrahimi, M. R. (2011). Detecting and investigating crime by means of data mining: A general crime matching framework. *Procedia Computer Science, 3*, 872–880. [Online].
Le-Khac, N., Markos, S., & Kechadi, M. T. (2016). A data mining-based solution for detecting suspicious money laundering cases in an investment bank. *CoRR*, vol. abs/1609.00990, 2016.
Nath, S. V. (2006). Crime pattern detection using data mining. In *Web intelligence and intelligent agent technology workshops, 2006. WI-IAT 2006 workshops. 2006 IEEE/WIC/ACM International Conference on*. IEEE, pp. 41–44.
Oh, K., & Cho, S. B. (2010). Semantic networks of mobile life-log for associative search based on activity theory. In B. T. Zhang & M. A. Orgun (Eds.), *PRICAI 2010: Trends in artificial intelligence* (pp. 643–648). Berlin: Springer.
Park, A. J., Tsang, H. H., & Brantingham, P. L. (2012). Dynalink: A framework for dynamic criminal network visualization. In *2012 European intelligence and security informatics conference*, pp. 217–224.
Qazi, N., & Wong, B. L. W. (2017). Behavioural tempo-spatial knowledge graph for crime matching through graph theory. In: *2017 European Intelligence and*

Security Informatics Conference (EISIC), pp 143–146, https://doi.org/10.1109/EISIC.2017.29.

Qazi, N., Wong, B. L. W., Kodagoda, N., & Adderley, R. (2016). Associative search through formal concept analysis in criminal intelligence analysis. In *2016 IEEE international conference on systems, Man, and Cybernetics (SMC)*, pp. 001 917–001 922.

Rasheed, A., & Wiil, U. K. (2015). A tool for analysis and visualization of criminal networks. In: *2015 17th UKSim-AMSS International Conference on Modelling and Simulation (UKSim)*, pp 97–102

Reich, B. J., & Porter, M. D. (2015). Partially supervised spatiotemporal clustering for burglary crime series identification. *Journal of the Royal Statistical Society: Series A (Statistics in Society), 178*(2), 465–480.

Schroeder, J., Xu, J., Chen, H., & Chau, M. (2007). Automated criminal link analysis based on domain knowledge: Research articles. *Journal of the American Society for Information Science and Technology, 58*(6), 842–855.

Stasko, J., G¨org, C., & Liu, Z. (2008). Jigsaw: Supporting investigative analysis through interactive visualization. *Information Visualization, 7*(2), 118–132.

Sun, Y. (2013). *Mining heterogeneous information networks.* 14, 20.

Takuya, W., & Masuhara, H. (2011). A spontaneous code recommendation tool based on associative search. In: *Proceedings of the 3rd International Workshop on search-driven development: Users, infrastructure, tools, and evaluation*, ACM, New York, NY, USA, SUITE '11, pp 17–20, https://doi.org/10.1145/1985429.1985434.

Thongsatapornwatana, U., & Chuenmanus, C. (2014). Suspect vehicle detection using vehicle reputation with association analysis concept. In *2014 IIAI 3rd International Conference on Advanced Applied Informatics.* pp. 436–440

Vural, M. S., & G¨ok, M. (2017). Criminal prediction using naive Bayes theory. *Neural Computing and Applications, 28*(9), 2581–2592.

Wong, B. W., & Kodagoda, N. (2016). How analysts think. *Proceedings of the Human Factors and Ergonomics Society Annual Meeting, 60*(1), 178–182.

CHAPTER 12

All-in-One Next-Generation Community Policing Solution Powered by Crowd-Sourcing, Data Analytics, and Decision Support: The INSPEC²T Case

Sofia Tsekeridou, George Leventakis, George Kokkinis, Elisavet Charalambous, Dimitris Miltiadou, Nikolaos Koutras, Dimitris Katsaros, Peter Leškovský, Leonidas Perlepes, Antonis Kostaridis, Fotis Kouretas, Felix Wentworth, and Kam Star

S. Tsekeridou (✉) · D. Miltiadou
INTRASOFT International S.A., Athens, Greece
e-mail: sofia.tsekeridou@intrasoft-intl.com; dimitris.miltiadou@intrasoft-intl.com

G. Leventakis
Civil Emergency Planning and Civil Protection, Hellenic Parliament, Athens, Greece
e-mail: g.leventakis@parliament.gr

G. Kokkinis
Center for Security Studies (KEMEA), Athens, Greece
e-mail: g.kokkinis@kemea-research.gr

B. Akhgar et al. (eds.), *Social Media Strategy in Policing*, Security Informatics and Law Enforcement,
https://doi.org/10.1007/978-3-030-22002-0_12

Introduction

Community Policing (CP) has brought a new era for effective problem-solving for mainly low priority situations by encouraging the collaboration of citizens with police. More often than not, however, the police are still using traditional approaches to respond to problems identified in community settings. The research literature (e.g. Greene 2000) suggests that police officers' conception of their roles and their attachment to police work are improving with the adoption of community and problem-oriented policing roles. Smart Policing and Community Policing are called to provide solutions approaching crime from drastically different perspectives. The former uses highly advanced technological means with drastic measures for the elimination of a criminal case in minimal time, while the latter focuses on breaking down the barriers between the police and citizens in an effort to build, in the long term, a culture of communication and transparency (Greene 2000).

Regardless, both policing approaches have faced and still do face issues: Smart Policing has been characterized by low to moderate effects on targeted crime (revealing inadequate effectiveness of crime prevention and

E. Charalambous · N. Koutras
Advanced Integrated Technology Solutions & Services Ltd, Nicosia, Cyprus
e-mail: lc@aditess.com; management@aditess.com

D. Katsaros
EXUS Software Ltd., London, UK
e-mail: d.katsaros@exus.co.uk

P. Leškovský
Vicomtech, San Sebastian, Spain
e-mail: pleskovsky@vicomtech.org

L. Perlepes · A. Kostaridis
SATWAYS Ltd., Athens, Greece
e-mail: l.perlepes@satways.net; a.kostaridis@satways.net

F. Kouretas
IMC Information Management & Communications S.A., Athens, Greece
e-mail: fkouretas@imc.com.gr

F. Wentworth · K. Star
PlayGen, London, UK
e-mail: felix@playgen.com; kam@playgen.com

enforcement tactics to allow for a significant difference in crime measures), while various studies (e.g. Greene 2000, 2007) suggest that community and problem-oriented policing have had modest impacts on community crime with larger impacts being noticed on the quality of interaction between the police and the public. It has been further evidenced that policing strategies involving citizens in the design and implementation of innovative tools have a positive effect on citizen satisfaction and trust in the police (Greene 2000). Thus, many law enforcement agencies (LEAs) are investing in innovative technologies enabling effective collaboration with citizens to respond proactively to public safety concerns as well as to engage them more actively in policing activities while at the same time increase their trust (see Chaps. 2, 4, and 6).

This trend has led to the emergence of a variety and diversity of ICT-enabled solutions for Community Policing for different usage contexts and geographical spans. The major focus of this chapter is to review technological solutions for Community Policing to assess their strengths and weaknesses and subsequently to present an all-in-one fully featured solution for next-generation Community Policing, resulting from the H2020 EU-funded INSPEC^{2}T Project (No 653749) (INSPEC^{2}T 2018), demonstrating that even more value could be brought around by Community Policing practices and tools, when the needs of all involved stakeholders are effectively addressed. INSPEC^{2}T combines the good qualities from both Smart and Community Policing approaches, utilizing advantages offered by both disciplines while trying to overcome their barriers.

The structure of the chapter is as follows: in section "Community Policing Systems, a study on existing CP solutions is undertaken along with an analysis of stakeholders' needs and influencing factors for an effective CP solution that would be adopted by the relevant critical mass. Section "INSPEC^{2}T: An All-in-One Next Generation Community Policing Solution" presents a next-generation all-in-one CP solution, resulting from the INSPEC^{2}T project (Leventakis et al. 2016; Leventakis and Kokkinis 2017; INSPEC^{2}T 2018), powered among others by a crowd-sourced data analytics engine and a special-purpose community platform that has been developed to meet the identified requirements and overcome the limitations of current CP solutions. Section "INSPEC^{2}T: Solution Deployment and User Feedback" presents results from the deployment and usage of the INSPEC^{2}T system by local stakeholder communities in five European cities, while section "Comparison with Existing Community Policing Systems"—based on the feedback received—outlines the key innovations of INSPEC^{2}T in comparison to current CP solutions and analyses its strengths and potential weaknesses to consider for effective market penetration. Section "Conclusion" outlines conclusions.

Community Policing Systems: A State-of-the-Art Review

Currently, available technological solutions for Community Policing (CP) are fragmented and customized for the needs of the specific place or stakeholder. An integrated end-to-end CP system, covering the entire communication and data processing lifecycle from citizen to police to citizen throughout a crime evolution is not available yet. A general-purpose social media platform is not ideal for Community Policing due to data protection and privacy concerns as well as the geographic extent. An indicative list of CP ICT-enabled solutions is given below emphasizing the pros and cons of each solution:

- ***Carbyne*** (Carbyne 2018) is a general-purpose, end-to-end ecosystem for emergency response for all relevant stakeholders and the public. The most related features to CP tasks are incident reporting and management, supported by live video streaming. The command and control centres are able to collect reports, prioritize them, and refer them to call takers who then dispatch rescue teams per case. *Carbyne* is not specifically focused on Community Policing and thus does not provide all relevant community-related communication and collaboration features as well as the backend required data analytics core.
- ***CitizenCOP app*** (CitizenCOP 2018) is a mobile app enabling citizens to voluntarily report a crime without disclosing their identity, secure themselves in a safe zone, access quick help, call police, define close groups, and receive notifications. The app is dedicated to CP and has been released for use in India, without being accompanied by a centralized solution to effectively manage and process incident reports and communities' discussions.
- ***Community on Patrol App*** (King 2016) is a mobile app for citizens, developed by Microsoft and Zco Corp. together with Miami-Dade County, USA, to allow citizens to collaborate with police, report crime, and its location and provide Miami-Dade's real-time crime centre with the data they need to prevent crime. The app is a special purpose one for the specific country, for mainly incident reporting.

- ***My Police Department (MyPD) App*** (MyPD 2018), used in the USA, allows easy and quick communication with the local law enforcement on a number of incident types with their location and related photos. It further enables receipt of alert notifications sent by the police and provides badges to users for active involvement. It is not accompanied by a centralized system/dashboard, in which all reports may be processed, visualized, and handled. Furthermore, its applicability is restricted to the USA, while interaction within communities is not supported.
- ***AlertCops*** (AlertCops 2018) is a citizen security alert mobile app that allows communication of citizens with Spanish law enforcement authorities in order to send geo-located alerts about a crime or security incident, whether they are victims or witnesses of such crimes. They are also able to chat with police and receive notifications. No backend centralized system exists, while again interaction at community level is not supported.
- ***Burgernet*** (Burgernet 2018) is an integral system, used by the Dutch National Police, for alerts targeted to citizens, police, and municipalities. Citizens use its mobile app, on which they may receive a voice or text message (or an e-mail in some cases) from the police according to their location, when the latter gets a report on, e.g. a burglary, robbery, or missing child, and wishes to request citizens to submit information. Citizens may then call the police at a dedicated number to provide information. In addition, citizens may be informed about an unsafe situation.

It is overall acknowledged that a fully fledged all-in-one solution providing tools and services for all stakeholders involved in CP activities is currently lacking. What is more, most existing solutions have a limited geographic use context and are not powered by a data analytics and decision support core that improves and speeds up the work of police.

Community Policing Requirements and Influencing Factors

Having reviewed an indicative list of known Community Policing ICT systems and contacted a critical mass of law enforcement agencies and police, academia, non-governmental organizations, municipality authorities, social workers, neighbour watch associations, and citizen communities in order to acquire knowledge on their needs and the extent that current solutions meet them, we have developed a summary of fundamental requirements in terms of CP system capabilities as well as in terms of effectively addressing societal, ethical, and legal concerns that need to be fully met by any promising ICT-enabled CP solution.

From a systemic point of view, crucial requirements in terms of expected CP system features and capabilities are categorized in the following main pillars:

- Effective integration of smart policing with Community Policing capabilities to accommodate the needs of all involved stakeholders
- Effective online communication and collaboration among all stakeholders
- Fully featured incident reporting with attached media from web and mobile clients
- Online communities' interaction and support, events planning, chatting, rating, and commenting
- Reduction of information overload, ability to filter and extract meaningful information from the bulk of crowd-sourced information in order to facilitate police work
- Automated statistics and summary reports generation
- Intuitive geo-localized information visualization and decision support
- Easy to use and learn tools, training capabilities through simulations
- Interoperation of all relevant tools and services within an all-in-one integrated system and legacy systems

In addition, societal challenges have been identified, linked to reluctance and time required for a cultural shift towards CP activities that pose further needs for effective means of awareness and education on Community Policing practices in local societies, targeted towards citizens' engagement and the building of trust (see also Chaps. 4 and 8).

From the ethical, legal, and societal impact point of view—considering that the developed solution offers services to the public and law enforcement agencies—adherence to the data protection directives is necessary, namely the EU General Data Protection Regulation (GDPR) applicable since May 2018 for data protection and the PCJA for the protection of personal data processed for the purpose of criminal law enforcement. Following those directives, it is necessary that citizens' data are used in compliance with the applicable terms of use, sensitive data are securely communicated and stored, preventing distribution to third parties without access rights for them, user personal information are not retained, etc.

Bearing the above needs in mind, the INSPEC²T CP solution, presented in the next section, has been developed to effectively address all of them, while overcoming limitations of existing CP solutions.

INSPEC²T: An All-in-One Next-Generation Community Policing Solution

The EU-funded H2020 INSPEC²T projects' scope was to develop a sustainable framework for Community Policing that effectively addresses and promotes seamless collaboration between the police and the community (INSPEC²T 2018). INSPEC²T was focused on a user-centric design and development approach and mobilized and engaged a critical user group mass, in the EU and abroad. Driven from accommodated transnational and multicultural best practices, it adhered to an approach where social, cultural, legal, and ethical dimensions are embedded into core user-centric design specifications and implementation procedures. The major innovations of the INSPEC²T solution are:

- To simplify and accelerate incident reporting on crimes or other citizens' concerns
- To enable communication among citizens and police/government by (1) a special-purpose social media platform for communities' interaction and (2) online communication and collaboration tools for all stakeholders involved
- To support effective decisions/response, using an information management and analytics engine and intuitive information visualization
- To provide an end-to-end fully featured CP solution for all involved stakeholders

In the above, the term "incident report" refers to a citizen report on an evolving situation that may require police intervention or respond to an appeal for information related to an incident. The term "incident" refers to a situation that has been validated by the police or trusted community members and now constitutes a case that the police needs to handle.

An overview of the INSPEC²T CP solution and its components is provided in Fig. 12.1. The police or other governmental authorities mainly use the Secure Portal (SP), the dashboard of all incoming information that invokes the data analytics engine to discover correlations and hidden insights from crowd-sourced data to facilitate effective decisions and timely response. Citizens use both the Public Portal (PP) and the mobile

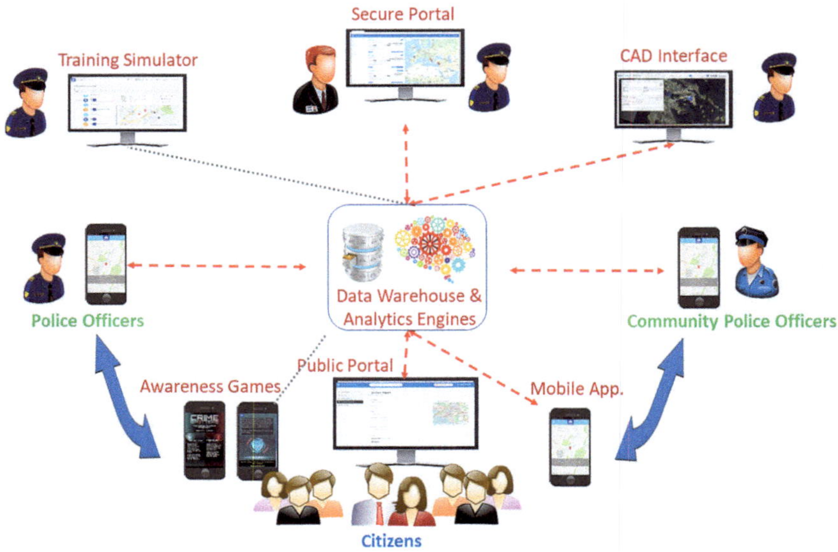

Fig. 12.1 The INSPEC²T solution at a glance (Source: Tsekeridou et al. 2018)

app to submit incident reports, interact in online communities, and actively help police work. Citizens communicate with CP officers through chatting, inbox messaging, calling, and texting. Citizens are further alerted on warnings and critical situations. A CAD (Computer-Aided Dispatch) Interface has also been developed to allow interfacing with legacy CAD systems, already in use by police. A Training Simulator has been developed to support simulations-based training of police users on INSPEC²T tools. Finally, mobile awareness games have been devised to engage and familiarize citizens with a Community Policing culture.

Crowd-sourced information comes in the form of:

- Incident reports from citizens (including textual, geographical, and media data), helping police work to resolve a disturbing situation faster and more effectively
- Posts and media shared in public or private communities, created in a special-purpose community platform, the Public Portal, and over the mobile app
- Inbox and chat messages from/to police, CP officers, other governmental authorities, citizens, etc. For example, municipal authorities may create a Municipality Community in their geographic jurisdiction, in which issues, events, or activities are discussed with citizens.

Alternatively, citizens residing in a neighbourhood may create a Neighbourhood Watch Community to discuss and be warned on issues in their vicinity.

- User ratings on police or other governmental authorities' performance and user engagement metrics measuring their active participation in CP activities online

The above assist in effective decision support, when coupled with intelligent analytics and intuitive information visualization. The INSPEC^{2}T Decision Support System (DSS) supports feedback provision in two layers: (1) during real-time information gathering a user can automatically gather intelligence for optimal decision-making at any given time, (2) during collection and storage of information distributed to stakeholders for improved strategic planning. Real-time information supports decision-making on a tactical level: where to dispatch police units, what kind of units to dispatch based on reported event, etc. On a strategic level, data correlations and business analytics enable automated generation of crime maps and crime trends along with intuitive information visualization.

With respect to compliance to PCJA and GDPR directives, a number of technological mechanisms ensuring data privacy and access rights have been implemented and put into practice. These involve the implementation of privacy by design components with well-defined user roles and data access rights, data integrity, ageing and erasure policies, stops of dissemination of irrelevant data as well as data minimization, proportionality, and portability.

Citizens–Police Communication and Collaboration

Citizens can communicate and collaborate with the police through incident reporting functions both in the PP and mobile app, in a registered or anonymous mode (see Fig. 12.2). Incident reports are structured, georeferenced, and time-stamped objects carrying structured and unstructured information in the form of text and multimedia objects (images, videos, audio). The contents of a report describe a situation that a citizen perceives as a safety issue or vulnerability that the police need to know about and act on. Incident reporting is done in a non-intrusive way to make the user feel comfortable about what they are to report; especially when the reporter is the victim herself. Thus, the user is allowed to file her report anonymously although she may have an active INSPEC^{2}T user account.

Through an easy to navigate interface, the user is called to provide information related to the type of incident that occurred, the time of the

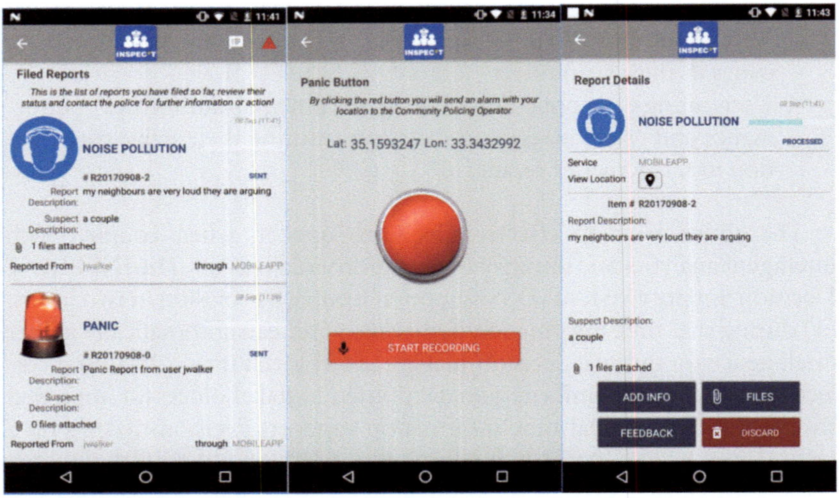

Fig. 12.2 Incident reporting capabilities

incident, its location as well as suspect and incident information. Additionally, the user is asked to indicate whether they are the victim of the reported incident, so that the operator is aware of this in case the user requires any additional support such as psychological support. Recording of multimedia data (audio, images, videos) is allowed to provide additional evidence to the police and are fed to the data analytics engine for further processing to provide important insights to the SP operator. A key feature of the INSPEC^{2}T mobile app is a panic-button-like functionality, which allows reporting of an incident in very few clicks transmitting the user's location and audio description.

Registered citizens on the PP or mobile app are encouraged to contribute valuable information, which contributes to their user participation score (Fig. 12.3 top), a 5-star rating that increments upon the ratio of the submitted citizen information that the police use for incident resolution. Citizens are further able to evaluate police work in resolving issues they reported (Fig. 12.3 bottom).

Police may communicate back with citizens via the SP to broadcast information, post-warnings in their communities, according to their geographic span or purpose, chat back reporting to registered citizens or CP officers under filed reports or managed incidents or share media and ongoing incidents (with or without files) to citizens, CP officers, and legacy systems (CAD and the Fire Services), requesting their help; when an

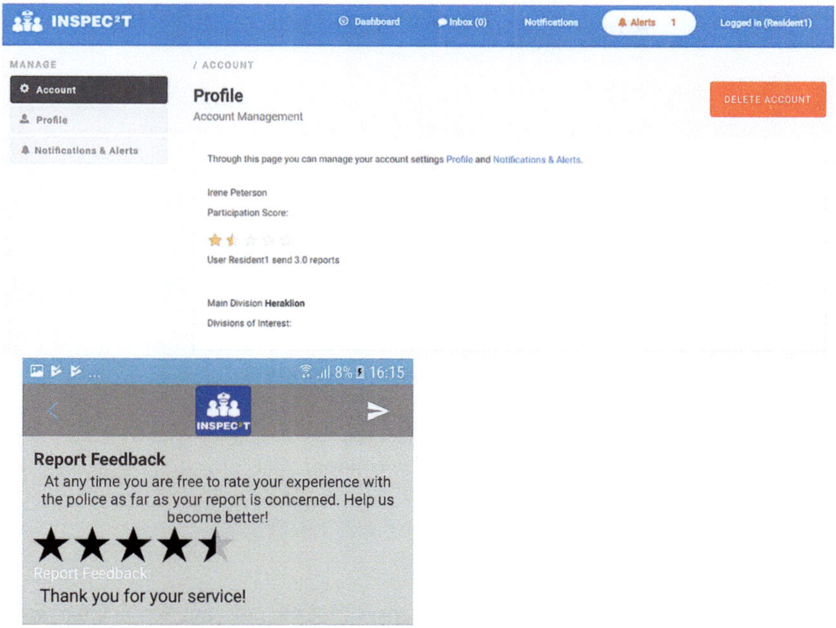

Fig. 12.3 User Participation Score at PP (top), Police work rating in the mobile app (bottom)

incident is broadcasted, users may submit updates to the system. Users are notified via push notifications. Upon incident completion, all involved users are automatically notified, and their access to incident information is erased for privacy and data protection purposes. Officer accounts are created only at the SP and marked "verified", i.e. validated accounts (Fig. 12.4).

A CP officer uses the mobile app for CPs, an enhanced version of the mobile app for citizens with additional functions for the efficient operation of policing actions. These are communication of report/incident-related content from the SP to CP officers, CP officer task assignment/acceptance for the investigation of incidents/reports as well as visualization of the exact crime location. Thus, the CP officer is further capable of managing incidents she is assigned to and providing updates to her status and chat back with the SP operator. CP officers assigned to an area are able to retrieve information related to their region or nearby regions according to their level of privileges.

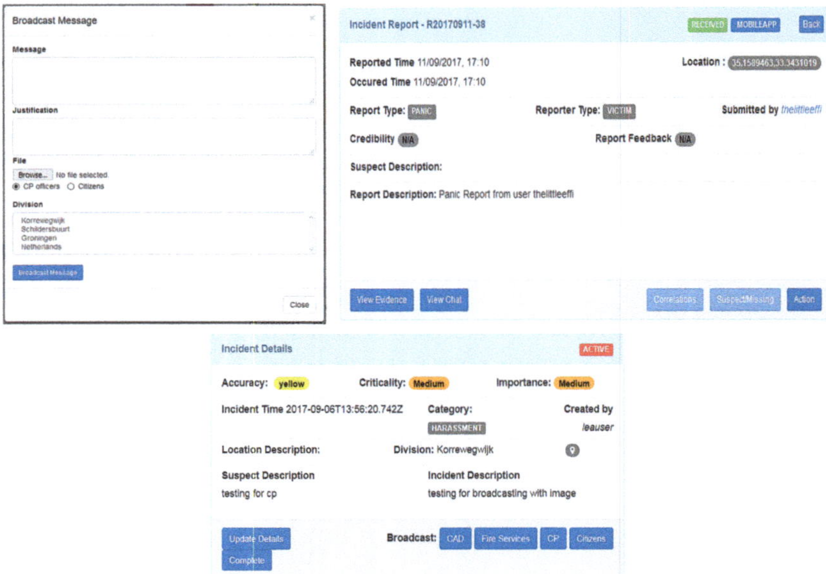

Fig. 12.4 Broadcast message to Citizens/CP Officers (top left), Chat with Reporting Citizen under submitted incident report (top right), Incident metrics and sharing abilities (bottom) at SP

What distinguishes the INSPEC²T tools for citizens (PP, mobile app) from other security-related apps is that it is not just for reporting, but it capitalizes on methods of increasing efficiency in report handling and allows immediate interaction and provision of feedback, thus resulting in improved and trusted communication and collaboration among citizens and LEAs. As a result, the LEA may easily trace a report in the system and provide feedback to the user. Similarly, the user will instantly get notified about progress updates on the reports they have filled, announcements about incidents that might influence the public as well as feedback about their participation. The aim is to make the user feel that their input is of value and that they form an integral part of the system.

With respect to security awareness, both the mobile app for CPs and the SP provide crime maps, in which the SP officer or a CP officer may retrieve occurrence rates of crimes per geographic division and their proximity to their location. Respectively, the PP and the mobile app for citizens provide a similar functionality for safe areas. However, since this feature is provided for raising awareness and not for panicking the public, the police in every instance is in control of what appears on the map.

Dashboard: Information Visualization, Workflow Management

The Secure Portal (SP) serves as the incident and data management portal of police to crowd-sourced information. The SP manages incoming citizens' reports and outgoing warnings and alerts in real time (Fig. 12.5). The primary SP user is the LEA operator, responsible for the management and handling of incoming information as well as the LEA data analyst responsible for the processing of generated analytics information. The SP provides functionalities on four main pillars: (1) real-time event management, (2) post-investigation analysis, (3) business intelligence analytics, and (4) user management. The SP operator manages incidents emerging from citizens' reports that are further correlated by the data analytics engine. Incoming reports and ongoing incidents are shown at its dashboard in a standardized list, while reports and incidents are mapped based on their geographic location. The SP map view page supports monitoring of human resources in the field, their assignment to incidents, and the presentations of nearby points of interest (POIs). The map view provides real-time monitoring of Community Policing officers' locations. The location of on duty officers is shown together with information related to

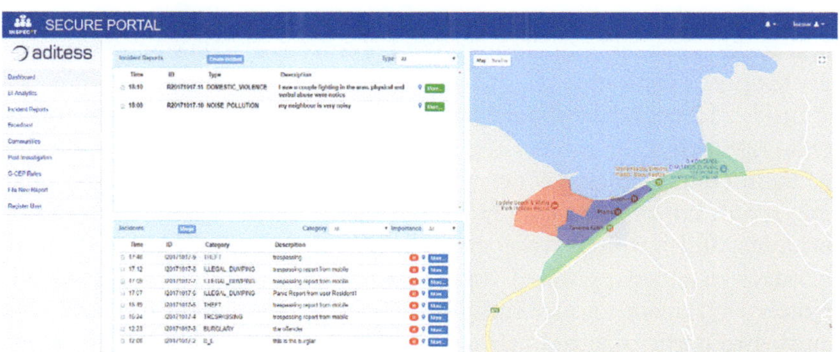

Fig. 12.5 Intuitive information visualization and real-time event management at SP: Reports on top, incidents at bottom, and map with divisions at right

their personal identifier; the locations of officers monitored by the CAD system (legacy system) are also visible.

INSPEC^{2}T resources are assigned to incidents based on availability and distance to their location. Officers assigned to an incident have access to all available information on their INSPEC^{2}T CP mobile application. For ease of management, prioritization of reports is supported based on their criticality, credibility/accuracy, and importance. *Criticality* refers to the intrinsic significance of the reported incident type. For example, a robbery is associated with a criticality level "medium", while noise pollution is associated with criticality level "low". *Credibility/ Accuracy* refers to a colour-coded value (traffic light scheme) related to the usefulness of information provided by the reporting source. Finally, *importance* is related to the urgency level assigned by the operator upon examination.

Citizens and Local Communities Interaction and Engagement

One of the main pillars of INSPEC^{2}T has been community engagement and increasing trustworthiness with LEAs through opening a direct communication channel, but most importantly through facilitating community building, bringing citizens of mutual interests and possibly concerns closer to allow them to collaboratively handle issues within their local societies. For the latter purpose, the PP—further from incident reporting functions—mainly comprises of a special-purpose community portal for citizens to engage in communities, either public or private, and to communicate with the police or discuss with other citizens on issues of their local society. Similarly, the mobile app for citizens attains similar community support functions.

In Fig. 12.6, the dashboard of the PP is illustrated, in addition to the menu options of the mobile app illustrating that citizens share posts and media in communities, plan events, and can submit incident reports and rate/report contributed content. Citizens receive alerts, shared media or incidents, and warnings from the police, while being able to send inbox messages to other registered citizens or the police. Citizens may opt to register or remain anonymous. By remaining anonymous, citizens are entitled to participate and provide information, but are excluding themselves from being contacted back. Registered citizens can participate in communities as well as express and exchange opinions in public or private

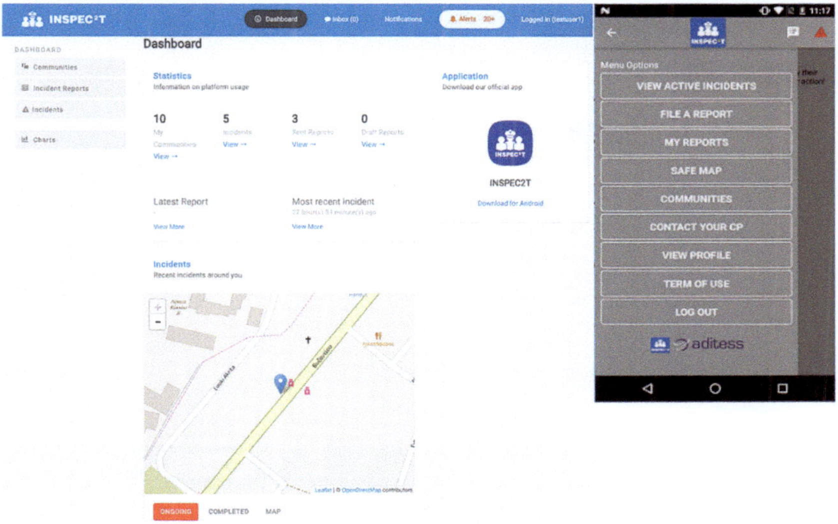

Fig. 12.6 Public Portal (left) and Mobile app (right) Dashboard and Menu, with communities' functions

discussions. The reason for devising a special-purpose community portal is that privacy and data protection aspects are better managed this way than in a general-purpose portal.

The notion of *community* in PP is generic and encompasses any organized act where multiple parties of common interests are involved. A community can be either "public", meaning that anyone can join without the need of approval, or "private", where new members need to be added by existing members. This generic concept covers both communities residing within the same geographical region and organized teams with activities influenced by the general community (i.e. taxi drivers, cycling teams, neighbourhood watch members). Community engagement includes also the ability of a citizen to directly contact their CP officer, either via phone, text, or email (on the mobile app) or through the inbox function (at the PP, see Fig. 12.7) and the CP officer's ability to broadcast message notifications to citizens registered under their community.

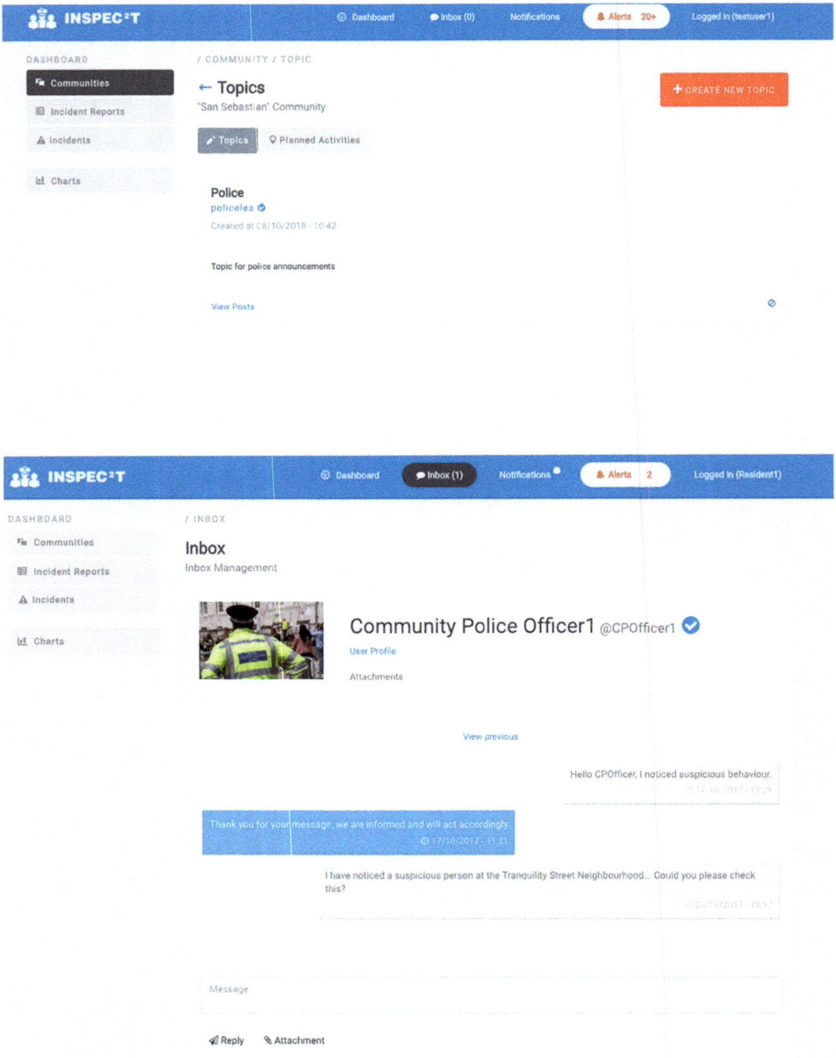

Fig. 12.7 Discussion Topics in Communities (top), Inbox message to a Registered User (bottom) at PP

Crowd-Sourced Secure Intelligent Information Management and Analysis

Further from the frontend tools operated by a variety of CP users, the INSPEC^{2}T system is significantly supported by an advanced data analytics and data management engine, handling crowd-sourced data in a secure and privacy-respectful manner. The backend secure data management and analytics engine presents a major differentiation in comparison to existing CP solutions, and thus a competitive advantage for the INSPEC^{2}T solution.

Upon the reception of a citizen report, its contents are automatically correlated with recently received reports spatially (location-wise), temporally (predefined temporal distance), semantically (in terms of textual descriptions of the event and potentially the suspect), and with respect to possibly attached multimedia content for suspect/missing person detection, image cross-correlation, and audio transcription. Correlation results are available for each citizen report at the SP and may assist the SP operator to reveal intrinsically hidden insights. Correlations can be manually approved or rejected by the SP operator.

Upon approval, a new incident is generated composed of the correlated reports, as shown in Fig. 12.8, in which a new incident on a "Burglary" in a neighbourhood has been created out of a number of correlated citizen reports shown below the map having been correlated spatio-temporally, semantically, and media-wise (see bottom pictures of submitted media data). In the latter case, face detection and cross-media correlations were discovered, while the detected face automatically matched the image of a criminal in the police database, reported at the SP dashboard.

In addition to near real-time analytics of incoming crowd-sourced data, INSPEC^{2}T further allows post-investigation functionalities in order to determine any hidden correlations and insights in stored crowd-sourced data over an extended spatio-temporal or crime-type span, determining further whether the same or a group of criminals have conducted a number of offences in different regions at different times. Post-investigation analysis, as shown in Fig. 12.9, allows the SP operator to query preserved data (that have not yet expired) and perform ad hoc data processing, revealing additional correlations with newly submitted information. The results of each post-investigation operation are logged and stored for future reference.

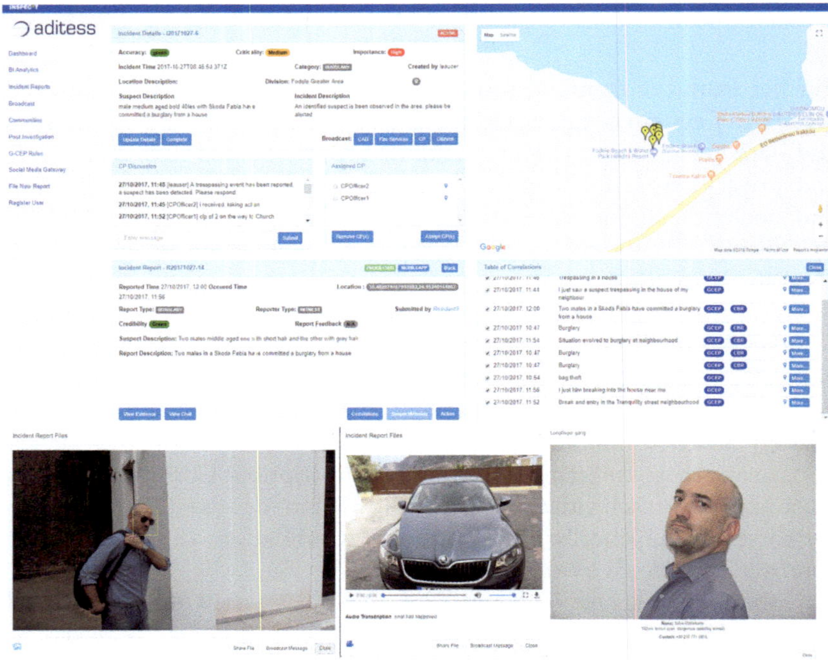

Fig. 12.8 Data analytics outputs in the form of spatio-temporal, semantic, and multimedia data correlations among incident reports at SP to facilitate police work and provide further evidence (The person shown was an actor of a simulated robbery scenario, who consented to take part in it and for his data to be used for the research purposes of INSPEC^{2}T.)

Spatio-temporal correlations among citizen reports are discovered by the Geospatial Complex Event Processing (G-CEP) engine that is responsible for correlating events in time and space and for providing actionable intelligence to LEA operators. It is recognized that, when dealing with multiple and disparate events, the Complex Event Processing (CEP) technology can deliver high-speed event processing, correlation, and identification. Users are able to define their custom correlation rules at the SP for both real-time and post-investigation correlation tasks, as shown in Fig. 12.10. These rules drive the discovery of spatio-temporal correlations among report-related events.

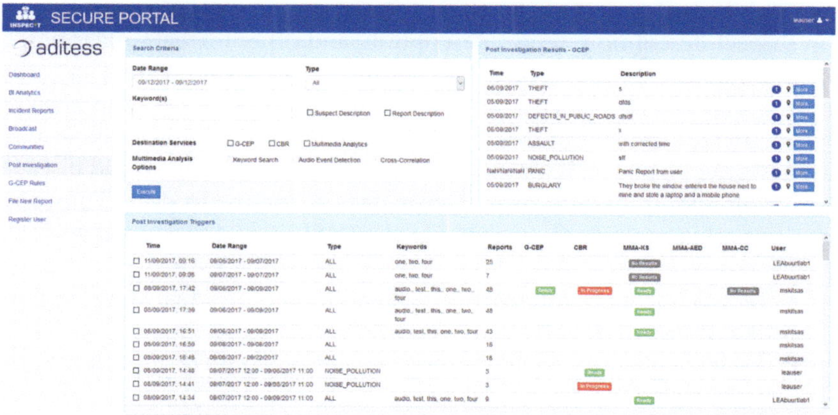

Fig. 12.9 Post-investigation Query Interface and data analytics results at SP

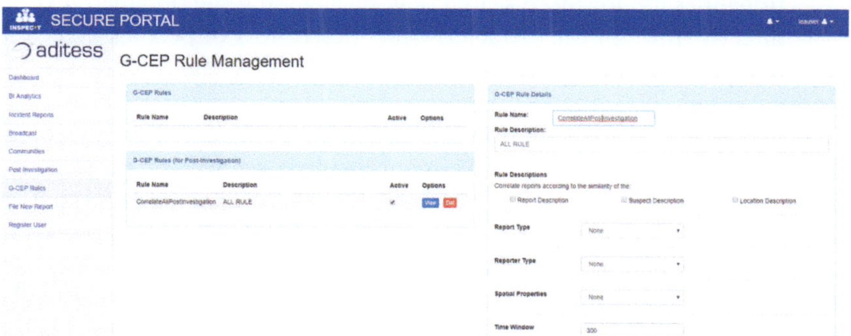

Fig. 12.10 G-CEP Rules definition at the SP

Citizens' reports are further correlated by Case Base Reasoning (CBR) algorithms that correlate reports according to their textual descriptions and semantics. Using semantic analysis, meaningful information is extracted from submitted reports that are matched among incoming reports, and thus semantic correlations are outputted. CBR's main goal is to assist with incident report grouping in real time, allowing the dispatch centres to act faster and smarter. Post-investigation, the search complexity is a significant trade-off for fast results, initiating algorithms to search through textual patterns in an effort to correlate past stored reports based on suspect description, report description, and other textual similarities.

Multimedia analytics (MMA) are further applied on media data, submitted as part of a citizen's report, to automatically process and analyse video, images, and audio data. Such data analytics tasks add value to CP, as they offer possibilities for more accurate information intake and discovery of significant insights and evidence data within media data. Thus, the goals of using multimedia analytics in the INSPEC^{2}T CP context are:

- To accelerate the information intake via speech to text transcription of audio streams
- To detect security-related acoustic events (e.g. gun shots, screams, or broken glass)
- To correlate images and videos based on person recognition, i.e. "the same person seen" rule
- And, finally, to index the uploaded content based on the above functionalities for advanced search (e.g. by analysing the obtained text or directly through the extracted visual descriptors of persons detected in the images or videos)

Different levels of multimedia processing are defined and are executed either automatically, i.e. upon incoming incident reports, or on request, i.e. as post-investigation initiated by the LEA. Audio data are processed in three modes: (1) non-speech segments are analysed to detect several acoustic events such as screaming, gun shots, broken glass, explosions, and siren sound, (2) speech segments are either automatically transcribed into text, or (3) searched for specific spoken terms, depending on their quality or the user's request. The latter two analytics tasks are currently implemented for UK English and Spanish. Although currently telephone speech is not processed within the INSPEC^{2}T platform due to its inherently lower quality, it is a feature of high interest for policing tasks applicable at the time of report intake since it considers the police–citizen communication via traditional means.

The person recognition analysis task of the INSPEC^{2}T system allows police operators to get automatic notifications if the same person gets identified in images or videos posted via different citizen reports. Besides automatically correlating the different reports, this feature is also useful for the re-identification of missing persons or suspect criminals by matching the detected persons from all received reports to an internal police database. Currently, the image analytics engine uses robust face and person detection models and deep neural networks to recognize the same actors based on face as well as whole person appearance similarity. Abstract appearance descriptors are used that do not reveal the real identity of the subject, respecting thus the relevant personal data protection issues.

Furthermore, data analytics results, with respect to usage data, provide useful insights about the impact of ICT-facilitated CP practices within local communities. Statistics data in the form of graphs, pie charts, and crime maps are thus generated by the INSPEC^{2}T system, both for the SP and PP as well as mobile apps, allowing trend detection not only over time or crime but also per geographical area. The analysed information is categorized as public and classified information according to who will have access to it. Public information can be publicly disseminated, usually approved by the LEA, e.g. statistics about how the system is used by the community over time, which communication technologies are preferred, and how many citizens actively participate (see Fig. 12.11).

Classified information refers to all types of available statistical information from different perspectives (time windows, location, type of crimes, geographical distribution, crime maps, and combinations of them), which are visible to only LEA users and CP officers to assist in their decision-making and strategy planning tasks. Figure 12.12 showcases an example of classified statistics and geo-located crime maps.

As a final consideration—but rather crucial for ICT-enabled CP solutions—secure and privacy-respectful data management in all lifecycle stages of the INSPEC^{2}T system has been implemented. INSPEC^{2}T's Data Warehouse (DWH) acts as a multi-faceted centralized data management system enforcing security on the data level and is responsible for the secure storage and management of INSPEC^{2}T generated data, while ensuring authorized and justified access to data. Data integrity of the original information is preserved with the computation of relevant hashes and digital signatures for any embedded multimedia object. The original data along with integrity preserving metrics are stored in a separate repository in which no user may perform changes. Additionally, and based on the imposed policy, data ageing attributes are attached to the object ensuring data retention only for the necessary and predefined period of time. The renewal of an item's date of expiration is possible and needs to be initiated from a user with the appropriate access key (separated from other functional permissions). DWH further implements mechanisms for maintaining extensive system logs and auditing. All incoming and outgoing

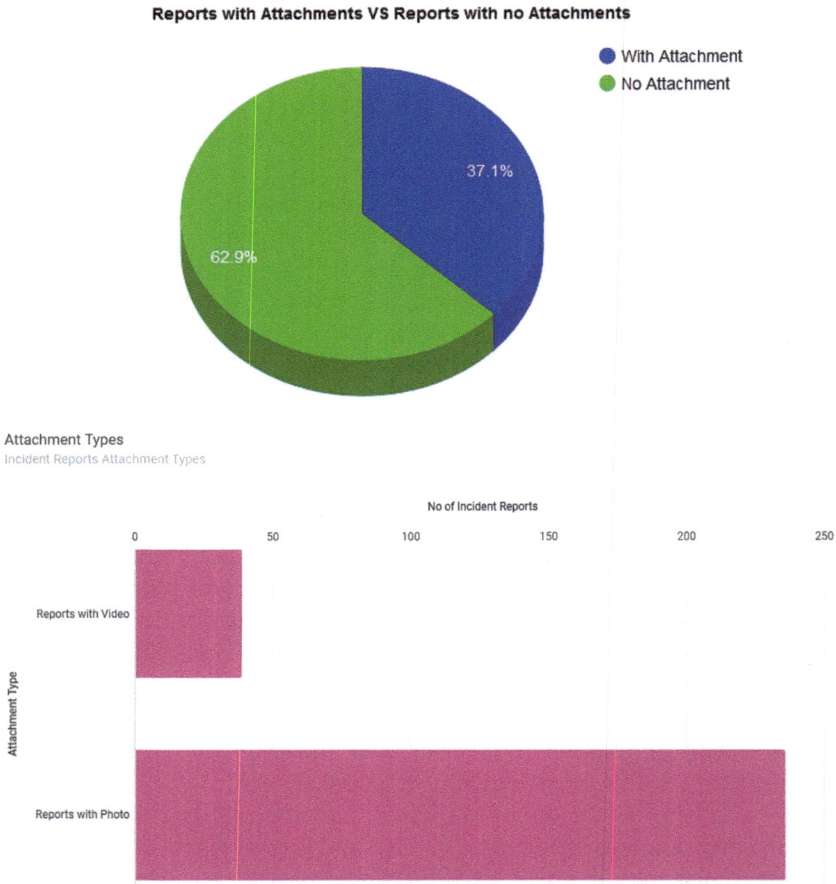

Fig. 12.11 Visualized public statistics on number of reports with or without attachments and the type of them

actions/requests are logged and retained in the system in the form of a record; only visible to users with appropriate permissions. Broadcasting and dissemination of data or files are only possible by the LEA operator on the secure portal; for the case of dissemination of multimedia files, the operator is also required to provide a justification of action.

The User Authentication Component (UAC) is responsible for user authentication and user account generation, determining which accounts

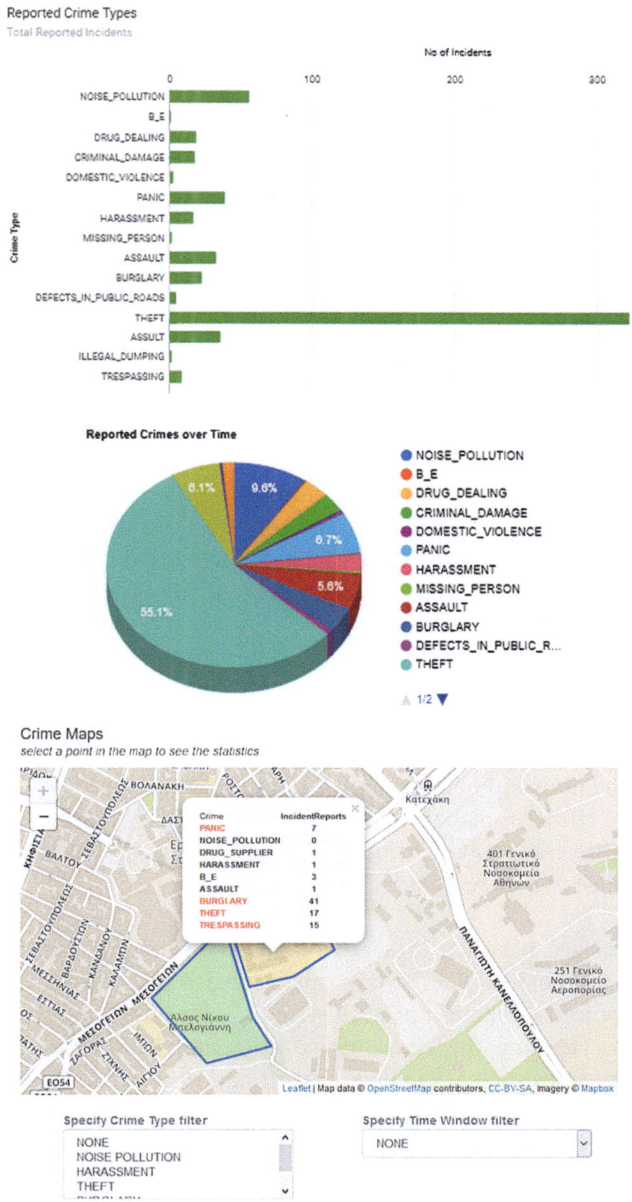

Fig. 12.12 Visualized classified statistics and crime maps for CP officers

are verified (e.g. police user accounts) or not (e.g. citizens accounts) as well as requiring only minimum user profile information while the user remains free to input any other personal information she desires.

Interface to Computer-Aided Dispatch Systems

Law enforcement agencies use CAD to facilitate incident response and communication in the field and support crime analysis (LEITSC 2005). Through years, CAD systems have evolved to support a much broader range of law enforcement strategies and tactics including problem-oriented policing and community-oriented policing strategies, which rely on a more proactive approach to policing (Greene 2007). According to research, less than 20% of citizen calls in a CAD system are for serious crime incidents (McEwen 2002). The rest are for incidents that affect the caller's quality of life to such an extent that they believe police intervention is necessary. The INSPEC^{2}T solution, in order to develop a sustainable and effective framework for Community Policing, comprises of an interface that enables information exchange between any CAD system and the INSPEC^{2}T system. The co-existence and integration of these two systems improve:

- The effectiveness of the LEA on the management of serious incidents by reducing the number of events/calls that the CAD system has to respond to.
- The relationship between police organizations and the communities; more efficient management of police forces will result in better responsiveness to Community Policing events, and as such the trust in police will increase.

The interface between these two systems is designed and developed according to the LEAs' everyday operational requirements in order to ensure optimum utilization of both systems. This interface is bidirectional in order to facilitate the exchange of operational information between these two systems such as incident reports and resource locations and statuses. Moreover, it enables the exchange of any Community Policing information between any CAD system and the INSPEC^{2}T's SP dashboard. As an example, a legacy CAD system (ENGAGE 2017) has been used to fully develop and test the CAD Interface, as shown in Fig. 12.13.

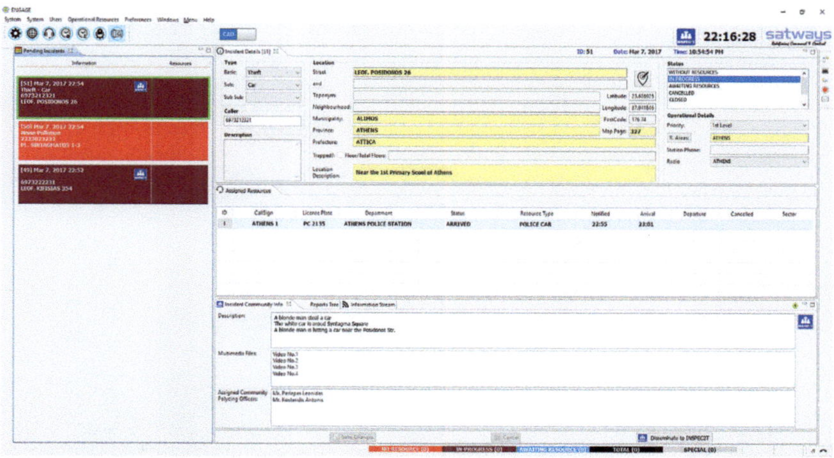

Fig. 12.13 ENGAGE IMS/CAD System interfaced with the INSPEC²T solution through the devised CAD Interface to share information and resources

Training Simulator

Simulation training is perceived as a very effective way for LEA operators to improve their critical decision-making skills. Training simulator sessions can be tailor-made to meet each individual trainee's needs allowing them to interact with the system, facing different situations or cases based on empirical scenarios with the ability to make errors, which in real situations could have fatal consequences. Equally important is the fact that simulation training is a very cost-effective alternative, allowing sessions to be repeated as often as necessary with different levels of difficulty and (close-to) realistic training conditions.

The INSPEC²T system further comprises of a scenario-driven training simulator to simulate the submission of citizens' reports and thus enable the training of police operators on the capabilities of the INSPEC²T system for effective management of simulated incidents. The training process is based on scenarios, created by trainers, and simulates real-life incidents. A scenario's timeline is a sequence of reports, delays, and signal-monitors, as shown in Fig. 12.14. The reports are injected in a training version of the INSPEC²T system, while delays may be added between reports in order to

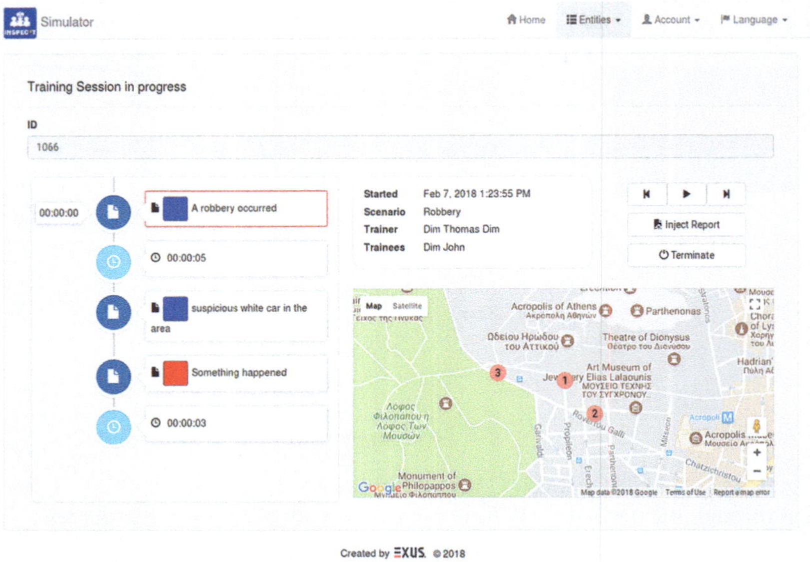

Fig. 12.14 Training Simulator—Scenario Execution: Incident Reporting Capabilities

simulate as accurately as feasible a real incident reporting phenomenon with submissions from the same or different citizens reacting to an incident. A scenario may also include signal-monitors, which are essentially delays in the execution of the scenario until a signal is transmitted by the INSPEC^{2}T platform. This functionality simulates the situation when LEA operators transmit an alert broadcast in order to notify registered INSPEC^{2}T citizens of an incident that has occurred in their vicinity, asking them for assistance. The simulator can intercept such a signal and resume a scenario's execution in order to simulate the expected increase of incoming citizen reports after an alert message is broadcasted. Trainers can customize the training sessions on a per trainee basis and even make ad hoc changes while a session is running. A trainer may pause and resume a session and move back and forth in the scenario's timeline. At the end of the session, the trainer scores the trainee's performance and provides feedback in the form of comments.

Awareness Game for Citizens' Familiarization with CP

In order to encourage citizens' engagement with CP activities, a mobile awareness game has been developed to allow citizens' familiarization during game play with CP practices. Games allow for an alternate approach to raising awareness to a particular problem or solution, targeting a wide market, and can do so either actively or passively. Games that both entertain and educate are a powerful tool to improve awareness and understanding. The decision to incorporate games to raise awareness of the INSPEC^{2}T project was to enforce the relationships and collaboration between police and younger members of the community, with the aim to improve the understanding of the role of citizens in supporting Community Policing and ultimately encourage players to download the INSPEC^{2}T mobile app and help in their community.

Resource Force is a single-player survival game, where a player must take control of her local police force (Fig. 12.15); it is her job to decide how to respond to incoming reports using limited officer resources, utilizing the help of citizens to provide more information or simply ignoring them, all while trying to keep the city happy and safe. The game is designed to

Fig. 12.15 Screenshot of the CP game *Resource Force*

challenge players, as they quickly discover that sending officers to all crimes will result in them having no means of resolving ongoing and developing crimes across the city. This results in players needing to think about their decisions and relying on citizens to help with providing more information and reducing required resources.

The survival aspect of the game improves engagement, helps maintain enjoyment and attention, while the "on-the-limit" game play creates an experience where players quickly realize that one wrong decision could lead to them losing entirely and regretting previous bad decisions. For each decision made in the game, players are given critical feedback, guiding them to make the correct decision in the future. The game was tested alongside the INSPEC^{2}T system to verify the effectiveness of using games as a method of raising awareness of CP and of the system itself. During testing with local students and members of the community, 84% of players indicated that they learnt something new about Community Policing, the reality of limited resources available or how the INSPEC^{2}T system works. To engage and continue to educate members of the community, it is important to add new content regularly to evolve with the shift in prevalent crimes across each city. To achieve this, the game is supported by an authoring tool, as shown in Fig. 12.16, providing a platform to create and manage new crimes in the game. The web tool is accessible to all, so anyone can propose new content to be added to their area, and once verified, it becomes instantly available within the game.

INSPEC^{2}T Solution Deployment and User Feedback

The INSPEC^{2}T system has been thoroughly tested by LEA stakeholders and local communities in five European cities (Belfast, Nicosia, Valencia, Preston, and Groningen) in a two-phase piloting and evaluation approach (Fig. 12.17). Initially, three pilots took place in pre-selected European cities. The phase 1 cities provided feedback and suggestions to improve the tested solution. The lessons learned together with identified best practices were used to plan the execution and assessment of the phase 2 pilots. Concurrently, system modifications and improvements have been introduced, while end-user requests for additional functionalities were addressed and identified system malfunctions overcome.

Four out of the five pilots have been conducted in controlled environment settings, with all participants having signed consent forms prior to their

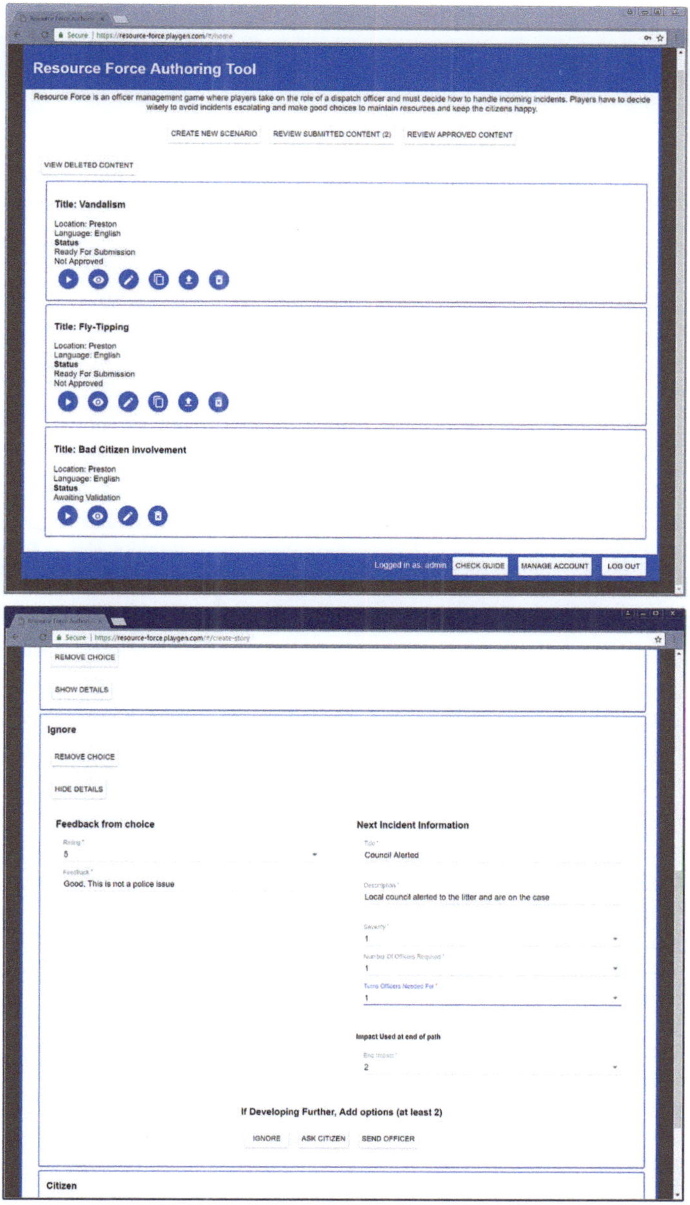

Fig. 12.16 Resource Force Authoring Tool dashboard and content creation page

Fig. 12.17 INSPEC²T Test case execution and assessment approach

participation in the tests, while the fifth was conducted in a non-controlled environment at the expense of disabling certain functionalities for safety purposes. All pilots were conducted in diverse urban environments, and the stakeholders from every pilot city had expressed in advance their expectations from the INSPEC²T system. Their expectations reflected their focus and their CP interests.

At the end of each pilot, a debriefing session was held with representatives from the involved user groups (citizens, organized community groups, NGOs, CP officers, city council officials, student associations, etc.). Using a structured questionnaire, their opinions about their experience with the system was sought. In addition, an online survey provided additional input for the overall experience of the end-users. From a technical viewpoint, the system proved mature and stable, while operating for a significant amount of time. From an operational point of view, the platform fits into all five CP operational models in Northern Ireland, Spain, Cyprus, England, and the Netherlands. Minor adjustments were required in order for the INSPEC²T solution to satisfy all policing demands and to be integrated in police command and control rooms. To summarize, great interest has been gathered by all relevant stakeholders, while the acceptance of citizens was very high. Throughout the period of planning and execution, no negative feedback was obtained. The stakeholders who had experienced the INSPEC²T solution asked to remain informed about the progress of the project and requested additional demos and pilots to take place.

Comparison with Existing Community Policing Systems

In Table 12.1, the INSPEC^{2}T Next-Generation CP solution presented in this chapter is compared against selected CP solutions. The comparison criteria have been defined in accordance with user and domain requirements and influencing factors, and thus focus on the CP system dimensions presented above. It is evidenced that the INSPEC^{2}T solution provides a fully fledged end-to-end modular CP solution (INSPEC^{2}T 2018) providing tools and services for all types of stakeholders involved in CP and featuring multiple capabilities that can easily be customized to any geographic area by diverse governmental stakeholders that benefit from crowd-sourced information, intelligence, and online communication and collaboration with citizens.

To further analyse the added value brought by INSPEC^{2}T in the respective market, a SWOT analysis was undertaken, the results of which are presented in Table 12.2.

INSPEC^{2}T is perceived as a competitive solution. Its main competitive advantage is that no other competitor offers all the listed strengths in one comprehensive solution. Moreover, INSPEC^{2}T offers many differences from existing CP solutions and a tailor-made alternative. Privacy by design and security by design have been pillars of its design process. Another advantage of INSPEC^{2}T is that the architecture is modular, which brings extra benefits for easier customization for local needs. In conclusion, INSPEC^{2}T does not have many competitors that offer the full package of services (cp. Table 12.1).

Conclusions

The INSPEC^{2}T system aims to combine the principles of CP with the affordances of new technologies. Both the actual and potential utilization of the resulting incident reporting and management tools (mobile apps, public portal) and crowd-sourcing processing modules (GCEP, MMA, CBR, BI) need to be compliant with legal frameworks, so that they can be correctly implemented in participating countries. This is the biggest challenge for a next-generation CP solution.

Following the conclusion of all five pilots, all involved stakeholders value the evolution of a CP solution to a social network system. The proposed solution allowed community members to get in touch with the

Table 12.1 Comparing existing CP solutions with the INSPEC^{2}T Next-Generation CP solution

CP solution	*End-to-end system*	*Incident reporting*	*Alert notifications and messages*	*Communities interaction*	*Data Analytics*	*Statistics/ crime maps*	*Interface with police CAD systems*	*Training simulation*	*Use purpose*	*Geographic usage*
Reporty	✓	✓		✓ *(family + friends)*					General purpose	IsraelUSA
CitizenCOP		✓	✓	✓ *(close groups)*					CP	India
Community on Patrol		✓	✓						CP	USA
MyPD		✓	✓						CP	USA
AlertCops		✓	✓						CP	Spain
Burgernet		✓	✓						Policing	Netherlands
INSPEC^{2}T	✓	✓	✓	✓	✓	✓	✓	✓	*CP*	*Any (with adequate customization)*

Table 12.2 SWOT analysis for the INSPEC^{2}T next-generation CP solution

Strengths	**Weaknesses**
• End-to-end CP system • Incident reporting • Alert notifications and messaging • Online communities' interaction and support • Data analytics • Statistics and crime maps • Interface with police–CAD systems • Training simulation • Customizable and modular • Privacy by design and security by design have been applied	• INSPEC^{2}T introduces new CP roles like the SP operator and administrator • A certain degree of ICT literacy and social media knowledge background is required • The introduced accountability (as citizens can monitor in real-time the progress of their submitted reports) necessitates adequate CP resources • Apart from human resources an organizational commitment to support CP is required
Opportunities	**Threats**
• Nowadays law enforcement agencies (LEAs) require end-to-end fully featured systems • Stakeholders require more sophisticated functionalities such as alert notification, incident reporting, decision support, intuitive visualization, online communication, and collaboration • LEAs welcome statistics and operational summaries on crime reporting • Stakeholders need training to use a multi-functional system • LEAs in any geographical area are looking into CP solutions to serve their local needs • Privacy by design and security by design approaches are a crucial part of LEAs' strategy	• Fragmented ICT modernization attempts resulted in a number of ICT systems that the INSPEC^{2}T solution needs to integrate (missing person's databases, police incident database, etc.) • The degree of commitment from LEAs both in terms of resources and budget to support a next-generation CP solution

police, report their problems, and observe online the police reaction. At the same time, citizens are able to create communities, discuss their security concerns and participate in discussions with other community members. All these functionalities are empowering citizens to join and actively participate in modern CP initiatives. In addition, an intelligent Community Policing system should be capable of supplying and feeding the existing police operational systems and, if needed, should have the capacity to provide facial recognition or offer acoustic event detection to alert operators.

INSPEC²T proved to be a successful proposition for a next-generation Community Policing solution. The developed solution, apart from the built-in innovation, is modular and customizable. The interface capacity and the extent to which INSPEC²T can be integrated with existing command and control systems will be key in determining its "marketability" to prospective end-users.

Acknowledgments The presented work has been conducted in the context of and fully funded by the European Union's H2020-FCT-2014 INSPEC²T project (grant agreement no 653749).

References

AlertCops. (2018). *Spanish Ministry of Interior* [online]. Retrieved from https://alertcops.ses.mir.es/mialertcops/en/index.html

Burgernet. (2018). *Dutch Netherlands police* [online]. Retrieved October 15, 2018, from https://www.burgernet.nl/

Carbyne. (2018). Carbyne Inc [online]. Retrieved October 15, 2018, from https://carbyne911.com/

CitizenCOP. (2018). *Citizen COP Foundation* [online]. Retrieved October 15, 2018, from http://www.citizencop.org

ENGAGE. (2017). *Incident Management and Computer Aided Dispatch (IMS/CAD) software system.* Athens: Satways Ltd. Retrieved from http://www.satways.net/products/engage-ims-cad/

Greene, J. R. (2000). Community policing in America: Changing the nature, structure, and function of the police. *National Criminal Justice Reference Service* [online]. Retrieved October 15, 2018, from https://www.ncjrs.gov/App/Publications/abstract.aspx?ID=185533

Greene, J. R. (2007). *The encyclopaedia of police science.* New York: Taylor & Francis.

INSPEC²T. (2018). *H2020 INSPEC2T Project Web site (Grant Agreement No: 653749)* [online]. Retrieved October 15, 2018, from http://inspec2t-project.eu/

King, J. (2016). *Community on Patrol App: Engage your community with app-based policing.* [Blog] Microsoft Industry Blogs. Retrieved October 15, 2018, from https://enterprise.microsoft.com/en-us/articles/industries/government/engage-your-community-with-app-based-policing/

LEITSC. (2005). Standard Functional Specifications for Law Enforcement Computer Aided Dispatch (CAD) Systems. Report of the Law Enforcement Information Technology Standards Council. Available online: https://it.ojp.gov/documents/LEITSC_Law_Enforcement_CAD_Systems.pdf

Leventakis, G., Kokkinis, G., Moustakidis, N., Papalexandratos, G., & Vasiliadou, I. (2016). Enhancing cooperation between LEAs and citizens: The INSPEC2T approach. World Academy of Science, Engineering and Technology. *International Journal of Social, Behavioral, Educational, Economic, Business and Industrial Engineering, 10*(8), 2836–2845. https://doi.org/10.1999/1307-6892/10005243

Leventakis, G., & Kokkinis, G. (2017). Enhancing cooperation between LEAs and citizens using social networks – The INSPEC T approach. Police Quarterly, 3(42), 64–66.

McEwen, T. (2002). *Computer aided dispatch in support of community policing.* Institute for Law and Justice [online]. Retrieved October 15, 2018, from http://www.ilj.org/publications/docs/CAD_Community_Policing_Exec_Summ.pdf

MyPD. (2018). *Wired blue* [online]. Retrieved October 15, 2018, from http://wiredblue.co/powerful-police-phone-app-for-community/

Tsekeridou, S., Leventakis, G., Kokkinis, G., Charalambous, E., Anson, S., & Sargsyan, G. (2018). A crowd-sourced intelligent information management and decision support system enabling diverse E-Government G2C2G. In: Proceedings of the Third International Congress on Information and Communication Technology (ICICT 2018), pp. 687–705.

CHAPTER 13

Multimedia Analysis in Police–Citizen Communication: Supporting Daily Policing Tasks

Peter Leškovský, Santiago Prieto, Aratz Puerto, Jorge García, Luis Unzueta, Nerea Aranjuelo, Haritz Arzelus, and Aitor Álvarez

INTRODUCTION

More than 50% of Europeans, aged from 16 to 74, benefit from communication over social media (Eurostat 2018). Social media provide them with the possibility of agile communication with individuals and groups. In the context of police–citizen communication, the social media channels play a vital role in strengthening the relations and collaboration between the police authorities and communities. A direct interaction with the citi-

P. Leškovský (✉) · S. Prieto · A. Puerto · J. García · L. Unzueta
N. Aranjuelo · H. Arzelus · A. Álvarez
Vicomtech, San Sebastian, Spain
e-mail: pleskovsky@vicomtech.org; sprieto@das-nano.com; apuerto@vicomtech.org; jgarciac@vicomtech.org; lunzueta@vicomtech.org; naranjuelo@vicomtech.org; harzelus@vicomtech.org; aalvarez@vicomtech.org

B. Akhgar et al. (eds.), *Social Media Strategy in Policing*,
Security Informatics and Law Enforcement,
https://doi.org/10.1007/978-3-030-22002-0_13

zens over such channels allows police to obtain valuable information that contributes to crime prevention and solving.

In this sense, social media is often used for distributing videos of petty crimes, violent acts and other public security related issues (often in private circles or publicly as a third party observer). Such material, if available, is nowadays collected and used by the police as evidence when resolving crimes. An example are the 2013 Boston Marathon bombings, where the police asked citizens to send photos and videos from the area in order to identify the suspects. An automatic processing of multimedia material may thus help in processing or searching for related evidence.

Nevertheless, the current multimedia sharing with police departments and its processing is limited. The available applications for community policing like *CitizenCOP* (CitizenCOP Foundation n.d.), *CommunityOnPatrol* (Miami-Dade County 2018), *MyPD* (WiredBlue n.d.), *myRPD* (The Reno Police Department n.d.) or *AlertCops* (Ministerio del Interior 2018) merely allow to send images, audios and/or videos but do not consider any automatic processing of the material.

In this chapter, the INSPEC^{2}T platform approach is presented, a tool for advanced police and citizen communication via modern information and communication technologies (ICT). The INSPEC^{2}T platform brings instantaneous safety issues and crime reporting via web access or mobile application to the citizens, and a platform for report intake and case management to the police (see also Chap. 12). Internally, it includes methods for automatic processing of multimedia contents, including images, audio and video. The aim of this platform is threefold:

- Simplify the information intake process at the time of reporting.
- Help in assessing the importance of each message.
- Help to handle the information overload of the police dispatch centre.

To this end, person detection and reidentification, automatic rich transcription and keyword search in spoken contents are the main selected technologies that should aid in achieving the above goals.

Multimedia Processing for Policing Tasks

Within the INSPEC^{2}T project and according to the stakeholders' preferences (Sargsyan and Stoter 2016), the following types of multimedia analysis were identified as of higher importance and value: audio speech to text transcription (e.g. for interviews or incident intake annotations), audio based security event detection, person detection and reidentification, action recognition (e.g. running, fight, burglary), security event detection (e.g. fire, smoke, panic) and entity specific object detection (e.g. car, policeman).

The final use case scenarios of the INSPEC^{2}T project envisioned that the citizens report events over their mobile phone by photographing, filming or audio-recording them. As this may limit the quality or comprehensibility of the input (e.g. for the cases of action and event recognition) three approaches were selected and implemented:

- Audio processing for rich transcription of spoken contents.
- Audio processing for keyword search of spoken contents.
- Person reidentification across multiple images or videos.

These technologies allow for agile text extraction from recorded speech (supporting goal 1 of the INSPEC^{2}T platform) necessary for indexing and search (goal 3), keyword search for automatic alarm raising (goal 2), person recognition for automatic notification of missing or suspect persons seen (goal 2) and automatic report correlation and grouping based on the "same person seen" rule (goal 3).

In the following sections, the developed technologies are explained in detail.

Audio Processing

Audio forensics involve the acquisition, analysis and evaluation of sound recordings that may ultimately be presented as admissible evidence in the law court or some other official venue (Maher 2009). This field includes many topics focused on the study of several automatic techniques for the processing of audio and spoken contents such as: establishing the authenticity of audio evidence (Koenig and Lacey 2015), performing speech enhancement to improve audibility and intelligibility (Ikram and Malik 2010), identifying the source device regardless of the speaker and speech content (Garcia-Romero and Espy-Wilson 2010), analyzing the acoustic environment to identify the place where the recording was made (Malik 2013), recognizing speakers (Campbell et al. 2009) and transcribing speech (Mattys et al. 2012). Among these, automatic speech transcription is probably the least applied technology in the forensic domain due to the artefacts involved (adverse acoustic conditions, low audio quality, spontaneous speech, overlapping, etc.), which makes it more difficult to obtain suitable results.

Nevertheless, in the field of community policing, where the citizens communicate incidents through voice calls, messaging (SMS, MMS) and other web and social media groups and frameworks, the analysis of spoken content becomes critical for the police, both at transcription and keyword-search level. The automatic transcription allows to obtain the complete spoken message in text format for further linguistic processing, whilst the keyword search serves as a useful technique for fast indexing and recovering purposes through the detection of specific words and key phrases in the audio.

This way, technology for automatic rich transcription and keyword search for English and Spanish languages was implemented and integrated in the INSPEC^{2}T platform, where the input audios were thought to be recorded by the user using the microphone of a mobile device at a short distance. This situation helped to improve the quality of the audio to be analyzed later. The technology developed used carefully trained deep neural networks for speech and language processing, taking advantage of the last modelling paradigms in the scientific community. *Rich transcription* involves the tasks of speech recognition, capitalization and punctuation. *Speech recognition* focuses only on segments containing speech and transcribes them to raw text. The *capitalization module* detects named entities and proper names and capitalizes them. Finally, the *punctuation module* adds full stops and commas to the capitalized text. On the other hand, the keyword-search technique exploits the generated lattice during the speech recognition process to seek for specific search terms given by the user. In order to guarantee a good robustness, the systems were trained with a great variety of acoustic and text data, ensuring a good performance of the acoustic and language models to adverse conditions.

Data Compilation for Model Training

The data compilation process involved the acquisition, transformation, processing and normalization of acoustic and text data to (1) train the acoustic and language models and generate the vocabulary of the speech recognition systems (rich transcription and keyword search), (2) build the capitalization components and (3) construct the punctuation modules. The data were collected for both English and Spanish languages.

Regarding the English language, the CSTR VCTK (Veaux et al. 2017) corpus was used for training and evaluation purposes. This corpus perfectly fits the needs of the INSPEC^{2}T platform, since it contains speech data (43 h) uttered by 109 native English speakers with various accents. With the aim of generating more training data and adapting the data to different acoustic environments, noise samples from restaurants, streets and shopping centres were collected from the Freesound website (Freesound Org n.d.), and mixed with the clean audios of the corpus. Thus, the amount of acoustic data for English summed up a total of 216 h. For the English text data, news from the crime domain were crawled from digital newspapers, totalling 47 million words (after normalization). This text corpus was then used to extract more data from a generic English text corpus using data selection techniques. This way, the final English text corpus was composed of a total of 92.8 million words.

The English capitalization module was trained using the final text described above (92.8 million words), whilst the English punctuation module was built using the same text and new acoustic data (234 h and 53 min) from already transcribed TED talks (TED Conferences n.d.).

Concerning the Spanish language, the SAVAS corpus (del Pozo et al. 2014) was used as a basis. It is composed of contents from the broadcast news domain, and the acoustic data is divided in clean speech (40 h) and noisy speech (100 h). Similar to the English language case, the clean speech was mixed with the noise samples, obtaining a total acoustic corpus of 220 h for acoustic modelling training. The Spanish text data was also obtained through data selection techniques, employing text from the crime domain gathered from digital newspapers (84.8 million words) and from the general domain. The final Spanish text was composed of 127.4 million words. Finally, the Spanish capitalization and punctuation modules were trained using the described acoustic (220 h) and text (127.4 million words) data.

Description of the System

The technology for automatic rich transcription and keyword search was constructed using the same tools and modelling paradigms for both English and Spanish languages. The only difference corresponded to the language-dependent grapheme-to-phoneme tool (G2P). For English, a statistical G2P tool trained with the Beep pronunciation dictionary (Hunt 1996) and the Sequitur G2P tool (Bisani and Ney 2008) were used, whilst a rule-based transcriptor inspired by the work of López (2004) was employed for Spanish.

The speech recognition systems for automatic rich transcription and keyword search were built using the Kaldi toolkit (Povey et al. 2011). The acoustic models corresponded to an hybrid deep unidirectional long short-term memory (LSTM) Hidden Markov Models (HMM) implementation, where LSTMs were trained to provide posterior probability estimates for the HMM states. Two types of language models (LM) were integrated per language: bigram Arpa-format LM for decoding and 5-gram constant Arpa-format LM for rescoring of the lattices. The decoding LMs were estimated with Kneser-Ney modified smoothing using the KenLM toolkit (Heafield 2011).

The capitalization modules were trained using the *recasing* tool provided by the Moses open-source toolkit (Koehn et al. 2007), whilst the punctuation modules were modelled as unidirectional LSTM Recurrent Neural Networks (Tilk and Alum 2015).

Finally, regarding the keyword-search technique, the lattices generated by the above described speech recognition systems were processed using the lattice indexing technique described in Can and Saraclar (2011).

Image and Video Processing

Main forensics video analysis currently operates on image level, working on video quality enhancements by sharpening, de-blurring, image stabilization (Amped 2018) and super-resolution (Bevilacqua et al. 2013); functionalities that offer higher quality images to the police investigators in order to proceed with further identification of objects, persons or actions. In the field of video surveillance, the focus has been on person detection and tracking as well as basic action recognition (e.g. persons entering restricted area, following a specific route, loitering, running, crouching or falling, BOSCH n.d.). Nevertheless, the current commercial solutions are tuned to specific use cases, scenarios and, most importantly, acquisition scenes that are usually well known a priori, considering that static CCTV cameras are used. Unrestricted processing of free-hand acquisitions and evaluation of complex events or actions is part of cutting edge research technologies. Scene understanding algorithms based on novel deep learning technologies show promising results in scene tagging and categorization (Zhou et al. 2014), event or action classification (Varol and Salah 2015) and object identification or dense image description (Johnson et al. 2016).

From all the topics, accurate person reidentification is currently one of the most required features that lately managed its breakthrough in the commercial sphere with applications that allow to process payments by verifying the identity of the customer by a selfie image (Petroff 2016) or tools for face identity management (Animetrics 2018) to law enforcement agents, corporations, night bars, etc. Considering the elevated maturity of these technologies, person reidentification was the primary implementation goal of the visual multimedia analysis within the INSPEC²T project. The envisioned use cases for the police case management system are: detecting missing or suspect persons, detect repetitive offenders and their typical whereabouts, check for possible implications of given offenders in the latest history of incidents, indirectly find a common witness of criminal or vandal acts and compile a concise list of persons present in given video.

To recognize the same person in images and videos, robust face and person detection models and open world metrics were employed. These capture the characteristics of an individual across different cameras, views, conditions and most of all generalize to an unlimited set of inputs. Face and whole body person recognition was considered. Global appearance measures were used that do not reveal the real identity of the subject, respecting thus the relevant personal data protection issues.

The principal steps of a general object reidentification include object detection, region normalization, feature extraction and final feature comparison for determining the identity of the object (Fig. 13.1). The detection task was solved by a classical machine learning approach by training deterministic classifiers on face and person regions. Currently, the Histogram Of Gradients (HOG) as well as convolutional neural network (CNN) features are used. The detection task involves a candidate region proposal that was based on a sliding window approach, run at multiple scales. Acceleration of the detection process can be obtained if the geometry of the scene is known, limiting the detection to certain regions and scales. In our case, the scale selection considered the minimal and maximal expected object sizes, set to a quarter of the image size for face detection. In case of CNN, a holistic approach was used, skipping the region proposal task, but including the region activation and evaluation within the layers of the neural network.

After faces or people are detected, they can be analyzed for reidentification purposes. Nevertheless, an indispensable part of the process is the image normalization. Besides colourspace normalization, the facial regions were extracted and rectified so that the faces became centralized and

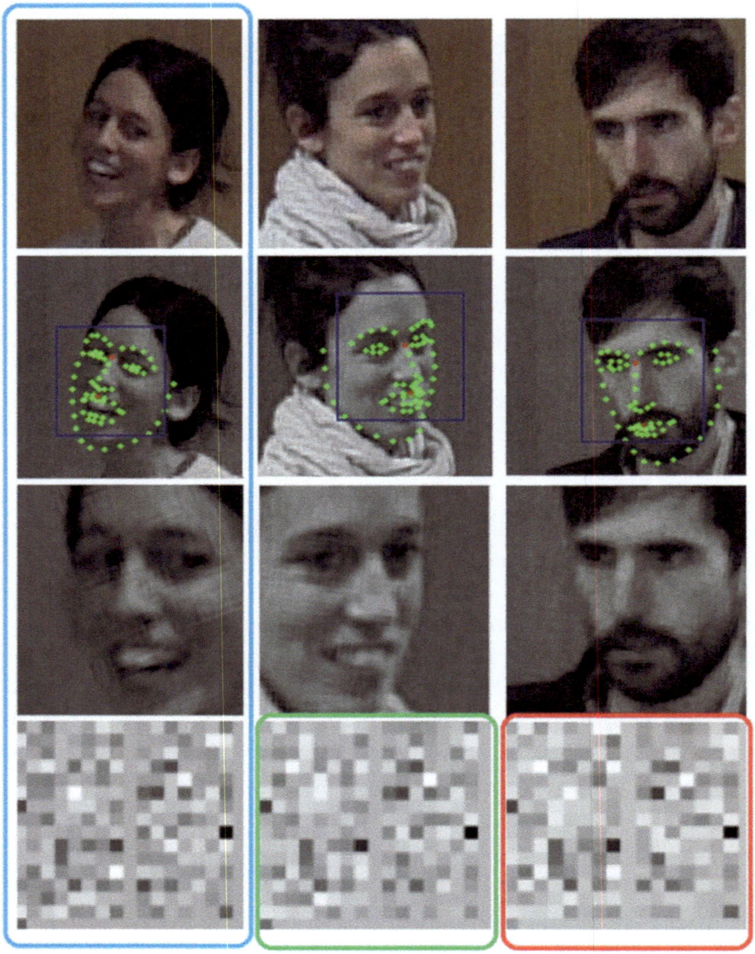

Fig. 13.1 Workflow of the person reidentification process based on facial images. The columns correspond to new person detection, matched person and no match, respectively. The rows show detected faces, facial landmarks detection, face region normalization via alignment and the extracted feature vectors (displayed as a matrix of grey values), respectively

aligned vertically. This process uses multiple facial landmarks (e.g. labial or eye corners) to estimate and correct for the inclination of the head. The detection of the facial landmarks used the method described in Kazemi and Sullivan (2014) and was pretrained using the iBUG 300-W dataset (Sagonas et al. 2013).

Finally, the reidentification task was based on the extraction of feature vectors, specific to each person or face, and its comparison in the multidimensional space. A 256-dimensional feature vector was extracted by a deep CNN for the face reidentification (Wu et al. 2015), whilst the popular histogram based features of 2784 dimensions (including colour and texture patterns obtained via a set of Gabor and Schmidt filterbank responses) were applied for the person reidentification, following the work in Zheng et al. (2016). The CNN features are preferred instead of handcrafted expert features, since in general they can better represent the input data and their characteristics (i.e. facial image) thanks to the high level of abstraction they consider.

For our purposes, a CNN was trained to extract the characteristic features that help to discern the facial images of two distinct people, thus maximizing their discriminability. The CNN was trained on the CASIA-WebFace database (Yi et al. 2014), consisting of images of more than 10,000 people with more than 15 images captured per person. For whole body reidentification, a distance measure was trained on a set of public datasets. Upon reidentification, the input image is compared with gallery images and the results are interpreted as a ranking list in which the first person of the ranking represents the gallery image with the most similarity. In order to output only one candidate, a distance threshold was adopted for face recognition and a binary distance metric trained through logistic regression for whole-body recognition.

Evaluation

The evaluation of the multimedia processing technologies has been based on publicly available databases as well as on domain-adopted data gathered from the system tests and first pilot executions. In the following the evaluation of audio and video processes will be described.

Audio Processing Evaluation

For the English language, the automatic speech recognition (ASR) and keyword-search (KWS) technologies were evaluated with a test partition of the CSTR VCTK corpus, which was composed of a total of 2 h and

48 min of synthetically generated noisy speech data. For the evaluation of the KWS system, 100 keywords related to the crime domain were manually selected from this corpus. The English capitalization and punctuation modules were evaluated over a test partition of the text compiled for the LM estimation (194 K words) and over contents from the TED corpus, which lasted 1 h and 31 min, respectively.

For Spanish, a test portion of the SAVAS corpus (1 h) was employed for the evaluation of the ASR, KWS and punctuation systems. Similar to the English language case, a set of 100 keywords close to the crime domain was selected for the KWS evaluation. The capitalization module was tested with a test partition of the text compiled for the Spanish LM estimation (159 K words).

The evaluations were performed using the Word Error Rate (WER) metric for the ASR, Actual Term-Weighted Value (ATWV) for the KWS, and the F1-Score measure for the capitalization and punctuation modules. The results for the punctuation module are presented separately for the period (full stop) and comma punctuation marks. Table 13.1 presents the performance for each system and language.

As can be seen in Table 13.1, the differences of results between both languages are noticable in terms of speech recognition and keyword-search detection. Even if the English corpus is more challenging given the different accents and acoustic environments involved, the ASR system performed very satisfactorily, with an error rate of only 4.87%. Besides, the KWS system based on the lattices generated by the recognition process reached an accuracy of 96.86%, which demonstrates that almost all the keywords were recovered from the audios. The performance of the Spanish ASR system was mainly influenced by the spontaneous speech included in the test data, which contains many street interviews recorded in adverse

Table 13.1 Performance of the INSPEC^{2}T audio systems using the WER, ATWV and F1-Score metrics for ASR, KWS and Capitalization/Punctuation, respectively

	ASR (%)	*KWS (%)*	*Capitalization (%)*	*Punctuation (%)*	
				Period	*Comma*
English	4.87	96.86	89.48	71.9	52.7
Spanish	18.07	75.45	97.08	70.2	48.9

acoustic conditions. However, more than three in four keywords (75.45%) were recovered from the lattices generated during decoding through the Spanish KWS system.

Regarding the Capitalization module, it was shown that the technological approach based on the SMT recasing model performed well for both languages. Finally, given that both Punctuation modules were evaluated with data containing spontaneous speech, the results reached through unidirectional LSTM RNN models can be considered interesting, even the ones obtained for the comma mark, usually influenced by the subjectivity of the person who employs it.

Image and Video Processing Evaluation

The face recognition task was tested and fine-tuned on the Labelled Faces in the Wild database (Huang et al. 2007), for four different distance measures: the Chebyshev (INF), Manhattan (L1), Euclidean (L2) and cosine (CS) norms. The results Fig. 13.2 demonstrate that the cosine norm

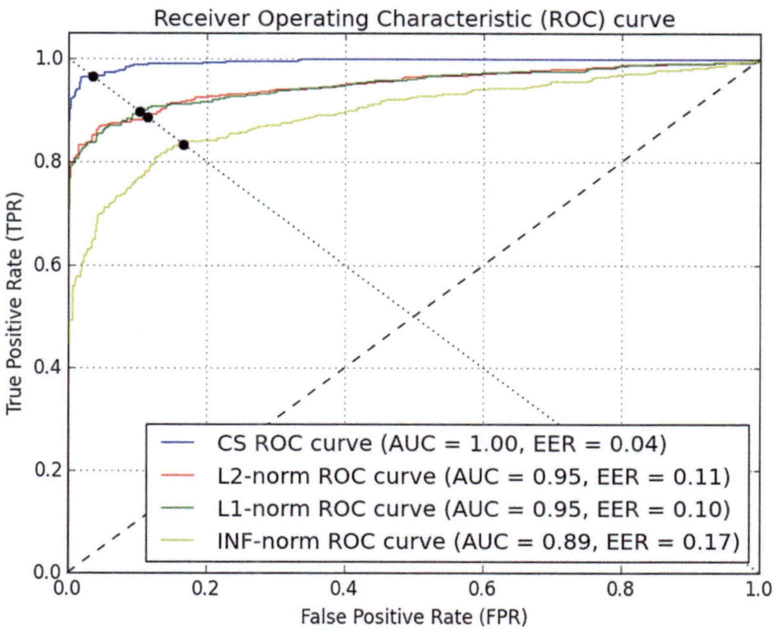

Fig. 13.2 ROC curve of the face reidentification model, considering several distance measures and evaluated on the LFW dataset

performed best. Consequently, the cosine norm with thresholds corresponding to equal error rate was used.

In practice, the reidentification model was trained on face images of resolution 120 × 120 pixels. In order to obtain such resolution on mid-range mobile phones,[1] the person would need to be at a maximum distance of about 6 m (considering an approximate face region height of 25 cm). Following internal experiments, solid performance was still obtained for lower resolutions, reaching 80 × 80 pixels. Such resolution allows to capture images of persons up to 9 m. Nevertheless, at these distances the image blur, caused by motion or out-of-focus during the acquisition, and low contrast, due to lighting conditions (e.g. in a garage), present more challenges and limitations for correct face detection and reidentification.

The detection task is the principal component that influences the correct run of the image and video analysis within the INSPEC^{2}T platform. For face detection four approaches, based on Cascade Classifiers, SVM-HOG, Max Margin Object Detection (MMOD) and CNN MMOD (Davis 2009), were tested on the face detection benchmark FDDB (Jain and Learned-Miller 2010). The results are given in Table 13.2.

Considering that the face reidentification module was limited to face images with a resolution of 80 × 80 pixels, the corresponding face detection results, evaluated on faces bigger than 80 × 80 pixels, are listed in column *Recall 80*. The best performance was given by the CNN approach.

Table 13.2 Performance of the face detection algorithms on the FDDB database

	Recall (%)	*Recall 80 (%)*	*False positives*	*Average runtime*
Cascade	62.5	75.9	102	1.47×
SVM-HOG	66.2	80.4	323	1×
MMOD	70.0	87.3	4	1.23×
MMOD CNN	72.7	91.0	2	22.35×[a]

The column Recall 80 refers to evaluation limited to face regions bigger than 80 × 80 pixels. The average runtime is expressed as a multiple of the fastest method. Discontinuous measure of the FDDB was applied

[a]The runtime of the MMOD CNN method, when run on GPU, was 1.25

[1]Google Nexus 5x was selected as a reference for the INSPEC^{2}T project; it includes an 8MP camera with a size of 1.4 micron pixels and 4.0 mm focal length.

Nevertheless, this is the most computationally expensive method that gives run-time performance only if GPU processing is applied. The traditional approaches, Cascade and SVM-HOG, as implemented in Itseez Inc (2015), show an unreasonably high false-positive rate. This problem can be contributed to a low precision of the underlying models. The best results, considering also the running time, were provided by the MMOD method as implemented within the public library (Davis 2009).

From the use case perspective, few problems have been experienced during the pilot execution for the face detection task: no false detections, and missed detection were mostly caused by the limitations on the model size, that is, the minimum resolution of the captured face and the blur of the image.

The person recognition task was tested on the VIPeR dataset (Gray and Tao 2008; Fig. 13.3). Nevertheless, three more datasets, the CAVIAR4ReID (Cheng et al. 2011), 3DPes (Baltieri et al. 2011) and GRID (Loy 2017), have been used for training to evaluate the cross-domain influence.

It can be observed that the ratios of the best person match are too low to be applied in practice. The results worsen for input data out of the training set, as would be the case for application within the INSPEC^{2}T platform. The main problems stem from vast backgrounds and pose variations.

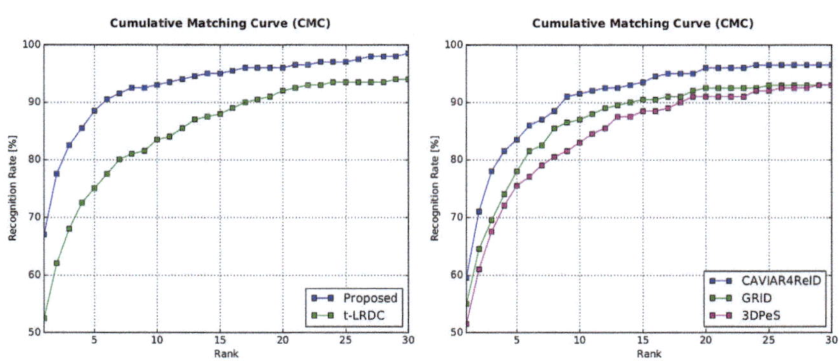

Fig. 13.3 Ranking curves of the person reidentification task in comparison to the method proposed in Zheng et al. (2016) (left) and when trained on additional datasets (right)

Conclusions

This chapter discusses multimedia analysis methods, applied within a ICT-based community policing platform, that help to automatize the analysis of audio and video files sent as evidence by the users to the police authorities.

The systems for audio analysis within this specific domain entails several difficulties related to (1) the variability in microphones and devices for audio capture, (2) the several adverse acoustic conditions, (3) the emotional state of the speakers or (4) the different accents, among others. In order to minimize their impact on the system's performance, multi-environment noisy synthetic data have been generated from corpora composed of utterances of speakers with various accents, besides using technology built with the latest modelling paradigms based on deep learning algorithms. Although the developed systems have achieved competitive results with the current test corpora, the future work will involve the acquisition of simulated recordings from pilots to evaluate the technology in nearly real environments.

The experience of the person reidentification functionalities within the first pilot executions (cp. Chap. 12) was positive, with little to no classification errors observed. Nevertheless, the real-world testing points on the necessity of the user's awareness on the favourable conditions and directives for capturing a video or a photo. The following are of major importance for obtaining the best results on the person reidentification task: (1) the camera should be held still, (2) the camera should have good focus on the captured subject, (3) the person's view with respect to the camera should be as frontal as possible, (4) the ambient illumination should be uniform and (5) the image resolution of the captured person should be of sufficient, preferably taken from within 9 m and with at least a 8MP camera resolution. Finally, the whole body appearance recognition system was not applied within the pilot tests due to its high false-positive ratio. In the future, additional image and pose normalization methods will be added in order to boost its performance and bring it closer to practical application.

Acknowledgements This work has been supported by the EU project INSPEC^{2}T under the H2020-FCT-2014 programme (GA 653749).

References

Animetrics. (2018). Advanced 2D-to-3D algorithms for face recognition applications. *Animetrics*. Retrieved October, 2018, from http://animetrics.com/forensicagps

Amped. (2018). *Amped Five*. Amped SRL. Retrieved October, 2018, from https://ampedsoftware.com/five.html

Baltieri, D., Vezzani, R., & Cucchiara, R. (2011). 3DPes: 3D People Dataset for Surveillance and Forensics. In *Proceedings of the 1st International ACM Workshop on Multimedia access to 3D Human Objects*, pp. 59–64

Bevilacqua, M., Roumy, A., Guillemot, C., & Marie-Line. A. M. (2013). Video super-resolution via sparse combinations of key-frame patches in a compression context. In: *30th Picture Coding Symposium (PCS)*

Bisani, M., & Ney, H. (2008). Joint-sequence models for grapheme-to-phoneme conversion. *J Speech communication, 50*(5), 434–451.

BOSCH. (n.d.). Video analytics at the edge. *Bosch Sicherheitssysteme GmbH*. Retrieved October, 2018, from https://ipp.boschsecurity.com/en/tools/video-tools/video-analytics-overview/intelligent-video-analysis-1

Campbell, J. P., Shen, W., Campbell, W. M., et al. (2009). Forensic speaker recognition. *J IEEE Signal Processing Magazine, 26*(2), 95.

Can, D., & Saraclar, M. (2011). Lattice indexing for spoken term detection. *J IEEE Transactions on Audio, Speech, and Language Processing, 19*(8), 2338–2347.

Cheng, D. S., Cristani, M., Stoppa, M., Bazzani, L., & Murino, V. (2011). Custom pictorial structures for re-identification. In: *British Machine Vision Conference (BMVC)*.

CitizenCOP Foundation. (n.d.). CitizenCOP APP. *CitizenCOP Foundation*. Retrieved October, 2018, from http://www.citizencop.org/?page_id=168

Davis, E. K. (2009). Dlib-ml: A machine learning toolkit. *J Machine Learning Research, 10*, 1755–1758.

del Pozo, A., Aliprandi, C., & Álvarez, A. Mendes, C., Neto, J., Paulo, S., Piccinini, N., Raffaelli, M. (2014) SAVAS: Collecting, annotating and sharing audiovisual language resources for automatic subtitling. In: *Ninth international conference on language resources and evaluation (LREC)*.

Eurostat. (2018). Individuals using the internet for participating in social networks, code: tin00127, Eurostat. Retrieved October, 2018, from http://ec.europa.eu/eurostat

Freesound Org. (n.d.). *Freesound*, Freesound Org. Retrieved October, 2018, from https://freesound.org/

Garcia-Romero, D., & Espy-Wilson, C. (2010). Speech forensics: Automatic acquisition device identification. *The Journal of the Acoustical Society of America, 127*(3), 2044–2044.

Gray, D., & Tao, H. (2008). Viewpoint invariant pedestrian recognition with an ensemble of localized features. In: *10th European Conference on Computer Vision (ECCV)*.

Heafield, K. (2011). KenLM: Faster and smaller language model queries. In: *Sixth workshop on statistical machine translation*. Association for Computational Linguistics.

Huang, G. B., Ramesh, M., Berg, T., & Learned-Miller, E. (2007). *Labeled faces in the wild: A database for studying face recognition in unconstrained environments*. Technical Report 07-49, University of Massachusetts.

Hunt, A. (1996). *BEEP dictionary*. Speech Applications Group, Sun Microsystems Laboratories. Retrieved October, 2018, from http://svr-www.eng.cam.ac.uk/comp.speech/Section1/Lexical/beep.html

Ikram, S., & Malik, H. (2010). Digital audio forensics using background noise. In: *IEEE International Conference on Multimedia and Expo (ICME)*.

Itseez Inc. (2015). *Open source computer vision library*. Retrieved from https://github.com/itseez/opencv

Jain, V., & Learned-Miller, E. (2010). *FDDB: A benchmark for face detection in unconstrained settings*. Technical report UM-CS-2010-009, University of Massachusetts.

Johnson, J., Karpathy, A., & Fei-Fei, L. (2016). DenseCap: Fully convolutional localization networks for dense captioning. In: *IEEE Conf. on Computer Vision and Pattern Recognition*.

Kazemi, V., & Sullivan, J. (2014). One Millisecond Face Alignment with an Ensemble of Regression Trees. In: *IEEE Conf. on Computer Vision and Pattern Recognition (CVPR)*.

Koehn, P., Hoang, H., & Birch, A., Callison-Burch, C., Federico, M., Bertoldi, N., Cowan, B., Shen, W., Moran, C., Zens, R., Dyer, C., Bojar, O., Constantin, A., Herbst, E. (2007) Moses: Open source toolkit for statistical machine translation. In: *45th annual meeting of the ACL on interactive poster and demonstration sessions, Association for Computational Linguistics*.

Koenig, B. E., & Lacey, D. S. (2015). Forensic authentication of digital audio and video files. In *Handbook of digital forensics of multimedia data and devices*, (pp. 133–181).

Loy CC (2017) QMUL underGround re-IDentification (GRID) dataset, School of Computer Science and Engineering, Nanyang Technological University, Singapore. Retrieved October, 2018, from http://personal.ie.cuhk.edu.hk/~ccloy/downloads_qmul_underground_reid.html

López Morràs, X. (2004). *Transcriptor fonético automático del español*. Retrieved October, 2018, from http://www.aucel.com/pln/

Maher, R. C. (2009). Audio forensic examination. *IEEE Signal Processing Magazine, 26*(2), 84–94.

Malik, H. (2013). Acoustic environment identification and its applications to audio forensics. *J IEEE Transactions on Information Forensics and Security, 8*(11), 1827–1837.

Mattys, S. L., Davis, M. H., Bradlow, A. R., & Scott, S. K. (2012). Speech recognition in adverse conditions: A review. *J Language and Cognitive Processes, 27*(7–8), 953–978.

Miami-Dade County. (2018). *Download the COP app*. Miami-Dade County. Retrieved October, 2018, from https://www8.miamidade.gov/global/service.page?Mduid_service=ser1508773998289190

Ministerio del Interior. (2018). AlertCops: Law Enforcement Agencies App, Ministerio del Interior Gobierno de España. Retrieved October, 2018, from https://alertcops.ses.mir.es/mialertcops/info/info.xhtml

Panayotov, V., Chen, G., Povey, D.,& Khudanpur, S. (2015). Librispeech: An ASR corpus based on public domain audio books.In: *IEEE international conference on acoustics, Speech and Signal Processing (ICASSP).*

Petroff, A. (2016). MasterCard launching selfie payments. Cable News Network. Retrieved October, 2018, from http://money.cnn.com/2016/02/22/technology/mastercard-selfie-pay-fingerprint-payments

Povey, D., Ghoshal, A., & Boulianne, G., et al. (2011). The Kaldi speech recognition toolkit. In: *IEEE workshop on automatic speech recognition and understanding (ASRU)*, IEEE Signal Processing Society

Sagonas, C., Tzimiropoulos, G., Zafeiriou, S., & Pantic, M. (2013). 300 Faces in-the-wild challenge: The first facial landmark localization challenge. In: *IEEE Intl Conf. On computer vision.*

Sargsyan, G., & Stoter, A. (2016). *D3.4 2nd SAG Meeting Report.* INSPEC[2]T consortum public deliverable

TED Conferences. (n.d.). *TED Ideas worth spreading.* TED Conferences. Retrieved October, 2018, from https://www.ted.com

The Reno Police Department. (n.d.). myRPD App. The Reno police department. Retrieved October, 2018, from https://www.renopd.com/myrpd

Tilk, O., & Alum, T. (2015). LSTM for punctuation restoration in speech transcripts. In: *16th annual Conf. Of the international speech communication association (INTERSPEECH).*

Varol, G., & Salah, A. A. (2015). Efficient large-scale action recognition in videos using extreme learning machines. *J Expert Systems with Applications, 42*(21), 8274.

Veaux, C., Yamagishi, J., & MacDonald, K., et al. (2017). *CSTR VCTK Corpus: English multi-speaker Corpus for CSTR voice cloning toolkit.* University of Edinburgh. The Centre for Speech Technology Research (CSTR).

WiredBlue. (n.d.). *My Police Deapartment App.* WiredBlue. Retrieved October, 2018, from http://mypdapp.com/

Wu, X., He, R., & Sun, Z. (2015). A lightened CNN for deep face representation. In: *CoRR* arXiv:1511.02683.

Yi, D., Lei, Z., Liao, S., & Li, S. Z. (2014). Learning face representation from scratch. In: *CoRR*. arXiv:1411.7923

Zheng, W. S., Gong, S., & Xiang, T. (2016). Towards open-world person re-identification by one-shot group-based verification. *IEEE Transactions on Pattern Analysis and Machine Intelligence, 38*(3), 591–606.

Zhou, B., Lapedriza, A., Xiao, J., Torralba, A., & Oliva, A. (2014). Learning deep features for scene recognition using places database. *Advances in neural information processing systems 27.*

Index

B. Akhgar et al. (eds.), *Social Media Strategy in Policing*,
Security Informatics and Law Enforcement,
https://doi.org/10.1007/978-3-030-22002-0

D

MIX
Papier aus verantwortungsvollen Quellen
Paper from responsible sources
FSC® C105338

If you have any concerns about our products,
you can contact us on
ProductSafety@springernature.com

In case Publisher is established outside the EU,
the EU authorized representative is:
Springer Nature Customer Service Center GmbH
Europaplatz 3, 69115 Heidelberg, Germany

Printed by Libri Plureos GmbH
in Hamburg, Germany